The Hello Girls

The Hello Girls

America's First Women Soldiers

Elizabeth Cobbs

Harvard University Press

Cambridge, Massachusetts
London, England

First Harvard University Press paperback edition, 2019

First printing

Frontispiece: Merle Egan was twenty-nine, unmarried, and
inspired to serve. After the war, she fought the U.S. Army
another six decades. Merle Egan, 1917, National Personnel
Records Center (NPRC), St. Louis, Missouri.

Library of Congress Cataloging-in-Publication Data

Names: Cobbs Hoffman, Elizabeth, author.
Title: The Hello Girls : America's first women soldiers / Elizabeth Cobbs.
Other titles: America's first women soldiers
Description: Cambridge, Massachusetts : Harvard University Press,
2017. | Includes bibliographical references and index.
Identifiers: LCCN 2016038111 | ISBN 9780674971479
(hardcover : alk. paper) | ISBN 9780674237438 (pbk.)
Subjects: LCSH: World War, 1914–1918—Communications. |
Telephone operators—United States—History—20th century. | World
War, 1914–1918—Participation, Female. | United States. Army. Signal
Corps—History—20th century. | United States. Army—Women—
History. | Women soldiers—United States—History—20th century. |
Women veterans—United States—History—20th century. | Women
soldiers—Legal status, laws, etc.—United States. | Sex discrimination
against women—United States—History—20th century. |
World War, 1914–1918—Regimental histories—United States. |
Women—Suffrage—United States—History—20th century.
Classification: LCC D639.T4 C63 2017 | DDC 940.4/173082—dc23
LC record available at https://lccn.loc.gov/2016038111

for

James Christopher Shelley

CONTENTS

Illustrations follow page 182.

Over there, over there,
Send the word, send the word over there—
That the Yanks are coming,
The Yanks are coming,
The drums rum-tumming
Everywhere.
So prepare, say a prayer,
Send the word, send the word to beware.
We'll be over, we're coming over,
And we won't come back till it's over
Over there.

. . . Hoist the flag and let her fly,
Yankee Doodle do or die.
Pack your little kit,
Show your grit, do your bit.
Yankee to the ranks,
From the towns and the tanks.
Make your mother proud of you,
And the old Red, White and Blue.

—GEORGE M. COHAN, 1917
(UNOFFICIAL ANTHEM OF THE U.S. EXPEDITIONARY
FORCES IN FRANCE)

PROLOGUE

S EVEN TOURING CARS, roadsters, and model T Fords constituted almost a traffic jam in the partially paved capital of windswept Montana in 1919. A local hero was coming
home. The wide western sky was still luminous that June
evening, when midsummer's light bathed the town well past
nine o'clock and crickets took up their nightly chorus late.

Some in the party that wheeled jauntily up to the train
station were veterans who had been discharged months earlier. Conspicuous with their colorful Victory Medals, the
men drew up into an honor guard and saluted as the tall,
dark-eyed woman stepped from the train in her blue army
uniform after crossing five thousand miles of ocean, forest,
and prairie. Thirty-year-old Merle Egan was a "big towner"
and one of the best long-distance telephone operators in the
United States even before she sailed to France. Now she was
a local celebrity with her fitted suit, brass insignia, tan parade gloves, and cocky aviation cap. Merle smiled and waved,
drained from her long ordeal but happy to spy the eager
crowd and the copper dome of the state capitol. It meant she
was home.

Montanans adored their girl, a soldier of "unique distinction" and "one of the few women who was in the military service of the United States," the *Helena Daily Independent*

boasted.[1] Many patriotic Americans had done their bit to win the Great War, but Merle Egan's bit was especially remarkable. She and two hundred other women had braved shot, shell, and submarines to operate the army's vital communications system overseas. General John Pershing recruited them personally. After the armistice, Merle commanded the switchboard for the epic peace conference at Versailles.

What wasn't clear as she shook hands with friends and admirers for nearly an hour on the crowded platform was where Merle Egan and other American women were going next. So much had changed in barely a year. The war had made women into soldiers and America into a world power. It had given the vote to women in Europe, then the United States. Only the week before, the U.S. Senate had approved the Susan B. Anthony Amendment after an exhausting seventy-year fight. Yet things were not entirely as they appeared. Despite the hero's welcome, Merle would soon discover, to her great surprise, that the army denied she had ever been a soldier.

This wasn't what officials promised when they first went looking for women to run the telephones that American generals needed in order to command every advance or retreat. And Merle Egan was a woman who believed in promises. Little did she know that evening in Helena that the next leg of her journey, from army switchboard operator to women's rights organizer, would take sixty years.

America's first female soldiers had been stationed throughout shell-shocked France as part of a compact branch of the U.S. Army known as the Signal Corps. The Signal Corps did not fire cannons, sink submarines, or bayonet invaders. Their job was to send messages. In March 1918, *Stars and Stripes* hailed the women who arrived during Germany's bombardment of

Paris as "Bilingual Wire Experts." They called them Uncle Sam's "Hello Girls."[2]

Army nurses also served in uniform. Yet theirs was an altruistic occupation designed to alleviate the ravages of war, not to advance military objectives per se. The purpose of Signal Corps operators was to help the United States win its war. They were soldiers, not angels.

Technology set the course, propelling change in unexpected ways. In May 1917, the month after Congress declared war, General John Pershing sailed for France on a ship stuffed with equipment. Nicknamed "Black Jack" for having commanded an all-black regiment on the American frontier in the 1890s, Pershing made sure to carry not only standard gear but also the newest devices.

Military tackle had undergone a revolution since the last Indian wars only two decades earlier, when Black Jack rode with the Buffalo Soldiers of the Tenth Cavalry. Planes had replaced horses. Trucks had overtaken mule trains. Telephone wires had outrun semaphore flags and smoke signals.

But equipment cannot fight. Pershing urgently needed people who could operate these machines instinctively. As a result, once Pershing and his generals arrived in Paris and observed the lay of the land, they ignored laws that said only men could serve in the army. The Industrial Revolution had called daughters and wives from the home to fill new jobs. Telephone operating was largely sex segregated. If America was going to position and command its immense forces, it needed women to handle the advanced technologies at which they were expert. They would have to withstand torpedoes, cannon fire, influenza, and petty-minded bureaucrats in order to send the word "over there."

Most worked behind the lines. A small group followed Pershing wherever he made his headquarters, from the short but intense Battle of St. Mihiel to the desperately drawn out Meuse-Argonne Offensive. They labored day and night within range of artillery that lit the horizon and shook their switchboards. This group achieved the highest aspiration of nearly every female Signal Corps member: to serve as near the battle as possible. Civilians may find it unconceivable that soldiers actually wanted to risk their lives, but volunteers accepted this as axiomatic. It's why they signed up.

The Hello Girls returned after helping to bring back two million doughboys, or infantrymen. Although their units were disbanded, they blazed the path for ongoing enlistment of women in the U.S. Army after 1943. Today, females constitute roughly 15 percent of the armed forces. Many aspects of their service seem routine rather than remarkable. Yet their role in combat remains deeply controversial. It challenges our beliefs about what females can or should do.

These same issues arose in World War I, when Signal Corps women first proved that remarkable acts could become routine. Afterward, the army refused to recognize women like Merle Egan as veterans. Some men turned their backs, while others saluted them. Revisiting this moment reminds us that institutions fight innovation. Individuals sometimes compel them to respond, but they usually fail. The experience of the Hello Girls is a microcosm of the ways that governments resisted sex-role change in the twentieth century—and into the twenty-first.

It also shows that change sometimes happens very quickly. In a short period of time, women demolished the barrier to voting that had long seemed insurmountable. Activists deserve much of the glory for amending the U.S. Constitution, but unanticipated events led listeners to hear old arguments in new ways. Technology had already expanded female contributions outside the home and the Great War demanded more. The convergence of feminism with these events gave women the vote and brought some into the U.S. Army. Then, when suffrage and the war were won, and women seemed poised to participate in society on a basis of full equality with men, the first wave of feminism dissipated like foam on the beach.

This wasn't altogether surprising. People find ways to preserve traditions they value while adjusting to changes they cannot avoid. Yet the most powerful trends are global, making them harder to resist. Emergencies hasten them.

The conflict into which the world tumbled almost by accident in 1914 altered expectations globally. Not only did the Russian, Ottoman, and German empires fragment into a dozen new nations, but cracks also ran under the British, French, and Dutch empires, as diverse peoples claimed a right to popular sovereignty. New republics proliferated: most unsteady, all imperfect. Subjects became citizens. Within older democracies, groups who had never had much of a voice raised theirs with new conviction.

Women used the conflict to achieve their long-standing demand for full citizenship. Australia, Canada, Denmark, Finland, Germany, Great Britain, New Zealand, Norway, Russia, Sweden, and ten other countries enfranchised females before America. The nation accustomed to congratulating itself as the vanguard of liberal democracy brought up the rear.

The United States was late to the war and women's suffrage. After President Woodrow Wilson endorsed both causes, he told the U.S. Senate that the popular vote was vital to the "realization of the objects for which the war is being fought."

He held out the hope that America might organize an enduring democratic peace. But how could it lead the free world if it was behind everyone else? Once entangled with foreign policy, women's suffrage suddenly became necessary, not discretionary. International credibility required it, and an aspiring world power must have credibility.

We know a great deal about the organizations that marched, sweated, and created a groundswell for female suffrage. We know less about how and why men changed their minds—and changing men's minds was critical, as only they had the legal authority to enfranchise women. More specifically, we have scant information to interpret President Wilson's most poetic lines to the resistant men of the Senate, who, with arms crossed, repeatedly rejected his pleas:

> This war could not have been fought, either by the other nations engaged or by America, if it had not been for the services of the women,—services rendered in every sphere,— not merely in the fields of effort in which we have been accustomed to see them work, but wherever men have worked and upon the very skirts and edges of the battle itself.[3]

The story of the women's Signal Corps unit allows us to look through the eyes of Woodrow Wilson and leaders in other democracies—new, old, and about to be demolished—and observe what they saw. The pioneers who served "upon the very skirts" of the battle help set a new standard of citizenship for women. They accepted the harshest responsibility of democracy: placing one's life in danger when necessary to defend the country. Their story, like a missing puzzle piece, completes the picture of how women around the globe not only demanded but also earned the vote.

Their victory, like most victories, was partial. The Hello Girls went to war at a time when women possessed citizenship only through their fathers and husbands. Merle Egan and her wartime buddies came back to a world in which females enjoyed new rights, yet some aspects of their status were mysteriously unchanged. This book calls them "soldiers" even though that label quickly became a source of contention. The U.S. government denied them bonuses, Victory Medals, honorable discharges, and a flag on their coffins. Although Signal Corps veterans embraced the whimsical moniker Hello Girls, they also wanted recognition as soldiers. And so they commenced a new struggle that eventually caught the second wave of feminism. Grace Banker, a twenty-five-year-old Barnard College graduate, led the Hello Girls to France. Decades later, Merle Egan finished their fight.

The Hello Girls explores how Americans mobilized for World War I, telephones transformed the United States, females joined the armed forces, suffragists won the vote, and women and men fought together for justice. It illuminates the battles that defined the twentieth century and still shape our own.

1

AMERICA'S LAST CITIZENS

MERLE EGAN'S full adventure spanned nearly one hundred years. The day she met with a newspaper reporter in August 1979, at age ninety-one, she wore the earrings explicitly forbidden when she served with the U.S. Signal Corps during the Great War. At that time, army rules prohibited any jewelry other than rings while in uniform, which meant no accessories except in one's quarters. It was the only place a uniform was not required. Even there, soldiers might wear civilian clothing only "when not receiving visitors."[1] So there was really no point to earrings. Aluminum dog tags on a string were the closest thing to personal decoration.

Although Merle told the reporter that she didn't care if the army gave her a proper ceremony, she certainly did. She had fought too hard, too long. Once leery of modern feminism, she now championed the National Organization for Women.

They were clip-on earrings, the sort her generation preferred. Merle wore them whenever she had her picture taken, such as when she spoke to elementary schools near the retirement home, telling the forgotten story to eager, upturned faces and showing them the doll with the smart blue uniform she had sewn.[2] She had attached the buttons from her own jacket and the metal insignia from the collar: a torch of gold between crossed wigwag flags above the initials "U.S."

The Signal Corps was a small branch of the army with its own motto—*Pro patria vigilans* (watchful for the country)—and song, its lyrics promising to "speed the message day or night."

The plastic doll wore no earrings, and its only concession to fashion were the black high heels painted on by the manufacturer. If Merle had had the talent, she would have cobbled a pair of regulation boots like the ones that had given her such trouble when she crossed the Atlantic on the *Aquitania*, praying a German torpedo would not send her and seven thousand doughboys to the bottom.

Now, compelled by Congress, the U.S. Army was at last ready to recognize her service. But was she ready for them? Why hadn't she accepted the army's invitation to go to Washington, D.C.? Perhaps she was still too mad at those who dared forget Grace Banker.

Merle turned to the reporter in the legal office of her friend and champion Mark Hough. "I'm surprised at them," she said. "Here they ignored us all these years, tried to pretend we didn't exist. Now they want to make a festival of it."[3]

Hough directed the conversation to a loftier plane. The secretary of the army, Clifford Alexander—the first African American to hold the position—might attend the ceremony. "It's not final yet," he said, but if Secretary Alexander came, Merle and "the relatives of those women who died before recognition was achieved" would be invited.

Merle broke into a smile and hugged the sweet-faced man, a boy to her at age thirty-four. "We couldn't have done it without him," she told the reporter.

And so she would wait. Mark Hough would call with the news. But it would be on an instrument very different from the receiver that President Woodrow Wilson had pressed to his ear when she connected him to Prime Minister David Lloyd George in 1919.

———————

At the outbreak of World War I, no one imagined its unintended consequences for women in the United States and elsewhere—certainly not the U.S. president.

In 1914, Woodrow Wilson opposed women's suffrage, the most dramatic of the demands activists had been making since the Seneca Falls Convention sixty-six years earlier. A whole generation had lived and died since Elizabeth Cady Stanton and Lucretia Mott called the assembly to discuss women's rights after finding themselves shunted behind a curtain at a London antislavery meeting to protect their modesty against their wishes. They had sailed thirty-five hundred miles to tackle the most important political issue of the day, only to be treated as an embarrassment.

Not that Woodrow Wilson wasn't progressive. He was a fervent Progressive who supported most innovations: the eight-hour workday, nature conservation, progressive income taxes, use of the secret ballot (known as the Australian ballot for its point of origin), and so on. In 1899, before his political career took off, Wilson wrote that government could improve America "by forbidding child labor, by supervising the sanitary conditions of factories, by limiting the employment of women in occupations hurtful to their health, by instituting official tests of the purity or the quality of goods sold, by limiting the hours of labor in certain trades, [and] by a hundred and one limitations on the power of unscrupulous or heartless men." Like many politicians, he paid close attention to the social reforms sweeping the country and rode the wave to power.[4]

Yet Wilson drew a line at votes for women, which violated the laws of nature, he thought. Female political activity was both offensive and ludicrous. It violated propriety. Only loose women, or old, used-up women, drew attention to themselves outside the home.

As a younger man, Wilson told his fiancée in 1884 about a Baltimore meeting touting the advancement of women. "Barring the chilled, scandalized feeling that always overcomes me when I see and hear women speak in public, I derived a good deal of whimsical delight . . . from the proceedings." One of the participants was "a severely dressed person from Boston, an old maid from the straitest sect of old maid." Indeed, the speaker was "a living example of—and lively commentary—of what might be done by giving men's places and duties to women."[5]

Wilson opposed female voting throughout his teaching at Bryn Mawr, a women's college whose students he found vacuous, and during his subsequent governorship of New Jersey. In his 1912 campaign for U.S. president, he told a colleague that he was "definitely and irreconcilably opposed to woman suffrage; woman's place was in the home, and the type of woman who took an active part in the suffrage agitation was totally abhorrent to him."[6]

Most men felt the same. So did many women, who constituted the backbone of antisuffrage organizations. The seclusion and subordination of females was centuries' old. In the same year that the Fifteenth Amendment prohibited disenfranchisement because of "race, color, or previous condition of servitude," a founder of the University of California commented that reforms like abolition rectified terrible historical abuses, but proposals such as women's suffrage trifled with biology. They courted disaster, theologian Horace Bushnell warned.[7]

Most nations never confronted the problem of enfranchising former slaves, but all wrestled with the vote for women. It was its own kind of world war, though largely nonviolent. Of the tiny handful of governments that allowed any democratic participation in the eighteenth and nineteenth centuries, none welcomed females. In enlightened England, prominent men vilified political women as "filthy witches" and "hyenas in petticoats."[8]

Female suffrage was so controversial that it was the only plank of the 1848 Seneca Falls "Declaration of Sentiments" that failed to pass the convention unanimously. Even Lucretia Mott, the elderly Quaker who called the gathering along with Elizabeth Cady Stanton, quailed at the possibly deleterious effects on the gentler sex: "Far be it from me to encourage women to vote or take an active part in politics in the present state of the government."[9]

Ridicule met those who dared. As the *New York Herald* described an 1853 meeting of suffragists, "The assemblage of rampant women which convened at the Tabernacle yesterday was an interesting phase in the comic history of the Nineteenth Century . . . a gathering of unsexed women, unsexed in mind, all of them publicly propounding the doctrine that they should be allowed to step out of their appropriate sphere." When a British suffragist chastised David Lloyd George in 1908—"We have waited for forty years!"—the leader of the Liberal Party replied to the amused crowd, "I must say the lady rather looks it."[10]

Even women sympathetic to the cause recoiled at suffragists who paraded in public. Few females wanted to be seen in the way Australian activist Louisa Lawson described the stereotype of a women's rights advocate in 1900: "an angular, hard-featured, withered creature, with a shrill, harsh voice, no pretense to comeliness and spectacles on the nose."[11] Few wished to risk the security and love that came from attracting a worthy husband.

In 1910, Dutch suffragists chose not to walk in their own parade, anxious to avoid "making a spectacle of themselves." Instead they pinned posters and banners to wagons and cars that male coachmen drove through the streets for them. In subsequent processions, as they became bolder, activists marched in demure folkloric costumes with white caps and wooden clogs to appeal to Dutch nationalism and cloak their controversial message in images of faithfulness, domesticity, and tradition.[12]

In countries under colonial rule, which described much of the world at the time, women walked an even higher tightrope. Egyptian women who wanted to be part of the independence movement had to contend first with the resistance of their own men. One of these women was fortunate to have a husband who read her statement aloud to a 1910 convention calling for Egyptian sovereignty, since she was not allowed to appear in mixed gatherings. Veiled women marched against British rule in March 1919, claiming their right to public space as well as an autonomous country. After independence in 1922, they demanded the vote. In 1923, feminists Huda Sha'rawi and Saiza Nabarawi uncovered their faces in an Egyptian train station as a political act.[13]

Acceptance of women's presence in public was grudging everywhere. When the inventor of the telephone, Alexander Graham Bell, told his fiancée, Mabel Gardiner, about Susan B. Anthony's speech outside Philadelphia's Independence Hall in 1876, she expressed dismay. Although Gardiner believed that women should be entitled to rights then widely denied, such as the privilege of owning property, Anthony's violation of "the public sentiment that forbids women to appear in public life," was unfortunate. To respectable women like Gardiner, activists willing to flout propriety attracted the stigma of fanaticism.[14]

Opponents of suffrage warned that feminism was un-American and would lead women astray. Ex-president Grover Cleveland told the *Ladies' Home Journal* in 1905 that the vote would have "a dangerous, undermining effect" on proper women. The suffrage movement, he said, was "so aggressive, and so extreme in its insistence, that those whom it has fully enlisted may well be considered as incorrigible."[15] The New York State Association Opposed to Women Suffrage warned in 1908 that civic life would be "disruptive of everything pertaining to home life." Women who rejected the vote embodied the highest values of "American Institutions, American Ideals, and American Homes," the New York organization claimed. It praised a kindred group in England: "The wiser women of Britain, the women opposed to the extension of the parliamentary franchise to women, are doing the right thing."[16]

Aside from the fear that politics desexed and degraded women, a central argument against political equality was that women were physically unequipped for a citizen's duties, particularly the obligation to defend one's country. Citizenship brought responsibilities with privileges.

As a British opponent of suffrage put it in 1907, "Women are quite as capable of expressing an opinion on political questions as men are, but they are not capable of enforcing it, they are physically disqualified."[17] In struggles involving contests of strength, women depended on men. If a nation could not oblige females to forfeit their autonomy, dress in uniform, and risk their lives, why should they enjoy equal privileges with those who did?

Bringing women into government also threatened to "sissify" nations at a time when international rivalries seemed to require virile men. Public leaders like Theodore Roosevelt cautioned against weaklings and pacifists whose fear of strife could rot a great nation "by inches." The turn of the century—an era when social Darwinists argued that only the fittest survived—was not the time to bring swooning damsels into government.[18]

"The primordial argument against giving woman the vote is that the vote would not represent physical force," a British author agreed in 1913. Enfranchising voters who had no capacity to defend the state endangered it. For "it is by physical force alone and by prestige—which represents physical force in the background, that a nation protects itself against foreign interference . . . and enforces its own laws." Chaos would result. Nothing could "more certainly lead to war and revolt than the decline of the military spirit and loss of prestige which would inevitably follow if man admitted woman into political co-partnership."[19]

In the United States, voting, fighting, and citizenship were braided together in the nation's origins. Popular sovereignty spread when propertyless Revolutionary War veterans demanded the vote. They had earned the franchise by aiming their muskets and risking their lives for the republic's defense. In the words of a toast raised in 1783 to celebrate triumph over Britain, "May all our Citizens be Soldiers, and all our Soldiers Citizens."[20]

Women were thus a lesser class of Americans. They couldn't be soldiers, and by law, they were an extension of their husbands and fathers under the rule of *feme covert*. Married women had no legal identity separate from their male protectors. Even their nationality wasn't all their own. They inherited their father's at birth and acquired their husband's upon marriage. Up through the 1950s, American-born brides risked their citizenship if they wed a foreigner. Some were declared aliens against their wishes.

In 1907, the U.S. Supreme Court rejected a California woman's appeal to keep her nationality after marriage with the observation that she chose "expatriation" when she chose her man. Suffragist Harriot Stanton Blatch, daughter of the famous Elizabeth Cady Stanton, lost her citizenship when she married an Englishman. Only his death in 1915 allowed her to reclaim it. Full, irrevocable citizenship was intertwined with self-sacrifice in war.[21]

Yet what if women volunteered to be soldiers? Would that make them real, true citizens, with all the appertaining rights?

Such questions were nonsensical until 1914, when the Great War became the first international conflict of the Industrial Revolution. Nations fought with equipment that rewrote physical requirements for soldiers, making some kinds of expertise as valuable as brute strength.

On the eve of this event, campaigners for women's suffrage had little to show for decades of unceasing effort. Leaders of the National American Woman Suffrage Association (NAWSA) had published innumerable pamphlets, circulated thousands of petitions, given countless stump speeches, and crisscrossed the country in buggies and trains with little observable effect. Harriot Stanton Blatch described the movement as stuck in "a rut worn deep and ever deeper."[22]

In contrast, the controversial Fifteenth Amendment enfranchising black men decades earlier had taken but a year to pass Congress and achieve ratification in 1870. (Though a civil war was required to make it thinkable, and Southern states soon subverted the law.) Nearly a dozen states granted the vote to immigrant aliens virtually without debate. America led the world in manhood suffrage, yet not a single state gave women this basic right of citizenship before 1890. By 1910, women had garnered the vote in only four out of forty-eight states, mostly former territories of the sparsely populated west that had previously enfranchised women.[23]

The legal treatment of women had otherwise improved, making resistance to the vote even more peculiar—and thus telling. By the last quarter of the nineteenth century, many of the "injuries and usurpations" condemned at Seneca Falls had been rectified. Thirty-three states and the District of Columbia had granted married women the right to keep their own wages. Divorced women had obtained shared custody of their children in most states, and wives could own land. Married women were finally allowed to make wills for separate property, though as late as 1940, a quarter of states still did not allow them to make contracts. The sexual exploitation of female minors had been mitigated, too. Age of consent laws that allowed men to wed girls as young as ten were raised to fourteen in most states, though in cases of incest some still required children to prove physical resistance or be judged "accomplices." Susan B. Anthony acknowledged in 1900 that changes in family law "represented a complete legal revolution during the past half century."[24] Since only males could vote, it was they who passed these laws—determined to do a better job protecting women, while denying females the power to protect themselves.

Yet paternalism was also matched by a new maternalism, or the notion that respectable middle-class women could take public stands on matters of "social housekeeping," like temperance and child labor, since such causes served others. Elite women such as Abigail Adams, Mary Todd Lincoln, and Edith

Bolling Wilson shaped the nation as well, through their public roles as the wives of politicians. Indeed, well-connected females had long been stereotyped as the "power behind the throne," expected to cajole and manipulate yet step back when commanded. But the idea of political power for oneself alone challenged the Judeo-Christian belief that God had plucked Eve from Adam's rib to be his helpmate. The vote epitomized personal autonomy. The highest privilege of citizenship simply did not apply to women, or so antisuffragists believed.[25]

The older regions of the country resisted change the most fiercely. Political machines in eastern states adamantly blocked women's enfranchisement. Southern states fought suffrage, too, opposed to anything that deepened the pool of black voters. And women themselves remained ambivalent, as revealed by an 1895 referendum in Massachusetts, when the Pilgrim state allowed both sexes to vote on a nonbinding resolution. The suffrage measure lost 187,000 to 110,000. Female opponents knocked on doors across Boston to defeat it.[26]

Activists lowered their sights. In the 1890s, the National American Woman Suffrage Association relinquished the aim of a national amendment like the one that enfranchised freed slaves. Instead, they refocused on state campaigns, most of which flopped. Public indifference increased. In 1900, eighty-year-old Susan B. Anthony turned over the presidency of NAWSA to her handpicked successor, Carrie Chapman Catt, a Wisconsin native anxious to breathe new life into the old-time religion.

A charming, relentless, masterful strategist, Catt served as president for four years, yet had to resign when her husband fell ill in 1904. His death the following year, combined with the deaths soon thereafter of her mother and brother, left Catt devastated. But she healed and took up the mantle again in

1915, replacing Anna Howard Shaw, a competent but less charismatic leader. A wealthy supporter of the cause died the same year, leaving Catt two million dollars and a treasure chest of emeralds, diamonds, rubies, and pearls. Catt put the gift in an independently managed trust—a war chest—for women's suffrage. The fight was back on. Membership rose from 117,000 in 1910 to two million by 1917.[27]

Catt hoped to gussy up the movement's image by recruiting "the best people," including heiresses, socialites, college students, and middle-class matrons. Whereas early reformers like Elizabeth Cady Stanton assumed women would enact radical social changes once empowered, Catt adopted neutrality on how women would use their votes or for which political party. She sought to put the tool in their hands rather than prescribe its use. This enabled suffragists to attract support across a broader political spectrum, the opposite ends of which had little in common. Catt believed suffragists should welcome support from any quarter. They would need it.[28]

The suffrage movement reached out to working women, too. The notion that women belonged in the home had been steadily undermined by industrialization. New jobs and personal need brought women into the labor force in record numbers. Working-class women did not have the luxury of staying home. They labored in textile mills, department stores, garment sweatshops, fish canneries, and manufacturing plants. Their participation in white-collared occupations swelled, too. In 1870, only 3 percent of clerks were female. By 1910, they numbered 35 percent.[29] Suffragists developed ties with the Women's Trade Union League, a cross-class coalition. "The disfranchised worker is always the lowest paid," the union declared in 1914, and in Boston, its leaders recruited eager telephone switchboard operators for the suffrage campaign. Accustomed to public parades to demand basic rights, unionists felt less constrained by gender conventions than did many other women.[30]

Carrie Chapman Catt also looked at the suffrage problem from a global perspective, which gave her hope. By the turn of the century, women's suffrage had begun to make progress on other continents, albeit slowly, and once again in thinly populated regions like Australia, New Zealand, and Finland. A founder of the International Woman Suffrage Alliance in Berlin in 1904, Catt informed the U.S. Congress in 1907, "The United States is by no means leading the world in the suffrage movement." She predicted that outside trends would pull America along. "Suffrage will ultimately triumph here as a result of its triumph in other countries."[31]

In 1916, when Woodrow Wilson won a second term as U.S. president, Catt announced an even more ambitious strategy, a "Winning Plan" to replace the decades-long campaign that had failed in state after state. The procedures for local constitutional amendments were too varied and cumbersome, she told followers. Although seven more western states had enfranchised women between 1911 and 1914, not a single polity east of the Mississippi River countenanced change. The movement needed to resume a two-pronged approach, aiming for a federal suffrage amendment while pushing local campaigns.

Southern members of the association balked. Racism and opposition to federal authority were joined like Siamese twins. Privately, Catt believed that America's "queer states rights sentiment" inhibited progress. Nonetheless, she prevailed. Southern feminists fruitlessly complained that they were "flattened by a well-oiled steamroller."[32]

Another complication emerged. Catt's resumption of the presidency coincided with a rebellion among some of NAWSA's younger members, who had lived in England. They turned toward militancy just as Catt was polishing NAWSA's respectability. Pointed at Parliament rather than diffused across a broad canvas, as in the United States, the women's suffrage movement in Britain was loud and uncompromising. "The suffrage campaign in the United States is a dull and commonplace affair when compared with the sizzling white heat of the British struggle," Catt admitted in 1912.[33]

Emmeline Pankhurst—the best-known leader of the British suffrage movement—advocated confrontation, though she combined incendiary rhetoric with refined manners and Victorian attire to foil aspersions against her womanliness. A scornful journalist dubbed militants *suffragettes,* but activists embraced the diminutive that connoted they were the epitome of femininity rather than its opposite. Pankhurst explained to Americans in 1913, "The extensions of the franchise to the men of my country have been preceded by very great violence. . . . Men got the vote because they were and could be violent. The women did not get it because they were constitutional and law-abiding." Pankhurst argued that even the most ladylike persons possessed the same capacity for physical force as men, and must display it to achieve full citizenship.[34]

Accompanied by acts of vandalism, such as setting fire to public mailboxes and breaking windows, Pankhurst's pugilistic rhetoric chilled Carrie Chapman Catt, as it did many British activists who were more conservative than Pankhurst. But a youthful rival in NAWSA, Alice Paul, felt otherwise. A Quaker descendant of William Penn, Paul was the brainy, brash, charismatic daughter of a lifelong American suffragist. She had earned her bachelor's degree at Swarthmore College, followed by a PhD at the University of Pennsylvania.

Alice Paul began her public career with an interest in social justice for the poor. It flamed into a passion for women's
suffrage during a three-year sojourn in England. Arrested
seven times and imprisoned three, Alice Paul bravely bore the
police brutality and humiliating forced feedings that met the
militant wing of the British campaign. But she did not go quietly, smashing windows and resisting arrest with the vim of
a Cornish coal miner.[35]

Philadelphia news-sheets relished reporting the unladylike
antics of the daughter of one of Pennsylvania's oldest families.
One came to Paul's defense, however, defending her reputation as a genteel woman thinking only of others. The young
lady was not a "vote-seeking rioter" but a "soft-voiced and
quiet mannered uplifter of the poor." Paul herself made no
such excuses. As she told the British magistrate who denounced her as "hysterical" before sentencing her to a month's
hard labor, "We feel that it is incumbent on every woman to
rebel against the state of political subjection in which we
are placed."[36]

Alice Paul returned to the United States in 1910, followed
by Lucy Burns, another American set ablaze by British tactics. Paul and Burns chafed against Catt's clever, but staid de-
corum. In 1913, with the approval of their elders, the younger
women organized a mass march in Washington on the eve of
President Wilson's first inaugural. The largest suffrage parade
America had ever seen attracted five thousand marchers and
a crowd of nearly half a million—including hostile gawkers
who taunted, spat upon, slapped, cursed, and threw burning
cigars at the peaceful protesters. The unwillingness of police
to protect the marchers further radicalized Alice Paul and
convinced many other suffragists that the moment for daring
measures had come. Around the time that Carrie Chapman

Catt became NAWSA president for the second time, Alice Paul and Lucy Burns established the National Woman's Party (NWP), a breakaway organization that eschewed state campaigns and aimed at nothing less than a federal amendment.[37]

Copying a British tactic, Alice Paul and her followers decided to hold elected politicians responsible. They focused their organizing on states where women possessed the vote and could punish recalcitrant politicians by electing better ones. The NWP scoffed at pandering to male egos. As Paul said, it was "more dignified of women to ask the vote of other women than to beg it of men."[38] In the elections of 1916, the party encouraged voters to eject Woodrow Wilson from office. When he won a second term despite them, they set up pickets outside the White House.

The generational and tactical split deepened. Susan B. Anthony's old organization followed a nonpartisan line. Carrie Chapman Catt wanted President Wilson's support. She denounced Alice Paul's aggressive tactics as a threat to suffragists' carefully laid plans to curry favor. The president considered the NWP more ridiculous than threatening, however, and tipped his hat sarcastically at ladies who held up signs whenever he exited his residence. Such women did not deserve the vote, Wilson believed. They confirmed all his worst opinions of feminists as unwomanly zealots.[39]

A Quaker, Paul gradually turned away from the militancy of the British. Instead, like Mohandas Gandhi during the same era, she advocated nonviolent civil disobedience. Paul walked the picket line after America entered the war in 1917, knowing that Washington police had been ordered to arrest suffragists for unlawful assembly and obstructing traffic. She received a seven-month jail sentence, called a hunger strike, and like other imprisoned women, endured painful forced feedings that caused vomiting and damaged their gums.[40]

Woodrow Wilson was unmoved. If anyone provoked his pity, it was those who assisted their country by performing jobs that strengthened the nation's development. By 1910, one-fifth of wage laborers were women, many in new occupations such as switchboard operating. Labor laws to protect them were one of Wilson's priorities. A good Progressive, he also took lively interest in reforms elsewhere. Wilson wrote his PhD on British parliamentary government, which he considered more effective than the congressional model. He told a campaign audience in Massachusetts during the 1912 election that the Scottish city of Glasgow was one of the "best governed in the world."[41] Americans could learn from others, he thought.

The proliferation of mass communication and transportation kept politicians like Wilson informed. It fostered a densely interwoven community of civic reformers on both sides of the Atlantic in the early twentieth century, including the younger suffragists whom Wilson disdained. Of all the new devices, none linked these reformers faster than the telephone—the invention that revolutionized warfare and allowed women to prove their fitness as citizen-soldiers.

2

NEUTRALITY DEFEATED, AND THE
TELEPHONE IN WAR AND PEACE

T HE WAR'S GLOBAL EFFECT on gender was in the future, but its ruinous consequences for France and Belgium commanded the world's attention in 1914. Over the course of three years, it eroded an American policy of neutrality dating from George Washington. The war also created an unforeseen but irresistible demand for female soldiers because of the simple telephone.

The Great War began not because a Serbian terrorist assassinated Archduke Franz Ferdinand and his wife, Sophie, when their automobile stalled on a side street of Sarajevo, but because of an alliance system in which ambitious rivals pledged to help special friends in times of trouble. Austria-Hungary was allied with neighboring Germany. The Kingdom of Serbia was allied with neighboring Russia, which was allied with France. All harbored ancient enmities that quickened suspicions. Once ignited, their feuds set the world on fire.

Austria began by attacking Serbia in July 1914 to punish it for harboring terrorists. This led Russia to mobilize troops on its border with Germany to thwart Kaiser Wilhelm II should he make a move to assist his ally. Germany responded by attacking Belgium, a neutral, largely unfortified country.

Indeed, Belgium had been inscribed onto the map nearly a century earlier almost entirely for its usefulness as an inviolate zone.

Ironically, the Belgians were a blend of French, Dutch, and German-speaking groups peaceably cohabitating. Most were Catholics whose sense of religious solidarity led them to break from the Protestant Netherlands after the chaotic Napoleonic Wars. The major powers recognized Belgium's independence in 1831 with the proviso that the new nation swear neutrality in all foreign conflicts and allow no one to use its territory as a springboard for attacking others.

A small, flat land, Belgium was a crossroads between France, Germany, and the Netherlands. The great powers pledged themselves not once but thrice to its neutrality: in the London Treaty of 1839, the 1870 treaty ending the Franco-Prussian War, and the Hague Convention of 1907. The German military manual at the start of World War I told junior officers that the borders of a neutral must never be crossed "even if the necessity of war should make such an attack desirable."[1]

German aristocrats did not apply such rules to themselves. The kaiser knew that if his Austro-Hungarian ally attacked Serbia, Serbia's Russian partner might attack Austria-Hungary. This might require Germany to defend Austria-Hungary against Russia, whereupon France would rush to the aid of Tsar Nicholas II. Plotting his way across the European chessboard, Kaiser Wilhelm II did not think he could defeat the Russian bear without first snapping the neck of the Gallic rooster. Paris could be conquered in six weeks, he was told.

France loomed as a threat because Germany had stolen two of its provinces forty-three years earlier. Alsace and Lorraine were hilly regions flanking the Vosges Mountains, a range of granite peaks, verdant meadows, alpine lakes, and

dense pine forests that provided a natural barrier against aggressive neighbors. Kaiser Wilhelm worried that France
wanted these provinces back. Of course, Germany brought
the problem on itself by annexing territories long integral to
the Gallic homeland.[2]

The Kaiser's next moves had been worked out years before
by Field Marshal Alfred von Schlieffen, safely dead when his
doomsday machine started whirring after the June 1914
murder of the Hapsburg heir. With Serbia under attack, the
Russian tsar positioned troops on the German border. This
prompted the kaiser to get a jump on France before Russia
could fully mobilize. Wilhelm II planned to attack Nicholas
II—who happened to be his first cousin—afterward. Respect
for Belgium's neutrality never figured into the kaiser's pre-
emptive calculations.

Europeans have debated for a century which nation must
bear primary blame for starting the war. At the peace conference that followed, England and France insisted that Germany
be judged guilty and handed an invoice for reparations. Some
argue this was unfair to Germany, which had reason to feel
encircled. It was the youngest of the European great powers,
unified under one flag in 1871 and mistrustful of Russian,
French, and English might. Placing the blame on Germany
challenged its honor and sparked Nazism.

Others refute this. The most influential have been German
historians like Fritz Fischer, who asserted that the empire was
just as aggressive as its enemies assumed.[3] Barbara Tuchman,
an American historian, paints a portrait of a thin-skinned,
pugnacious kaiser determined to make his cousins among the
European aristocracy respect him for once. That they already
respected German industrialization and social welfare systems made no dent in his insecurities.

This debate may never be resolved. What is undeniable is that Germany was first to strike a country with which it did not have a single grievance.

Not surprisingly, when the kaiser's troops invaded Belgium on their way to France in August 1914, England's ambassador demanded an explanation. Britain had repeatedly committed itself to Belgian neutrality. German chancellor Theobald von Bethmann-Hollweg protested that he could not fathom England going to war with a friendly, "kindred nation"—ruled by a grandson of Queen Victoria, for goodness sake—just to fulfill promises made on a "scrap of paper."

This was a feeble, foolish argument for a man who started a war in the name of German treaty commitments to Austria. Worse, the chancellor had not accounted for the English attitude toward promises made nor Britain's desire to maintain the neutral buffer across the channel from its own shores. The "scrap of paper" became a rallying cry for indignant Britons (as well as Americans) during four appalling years of war. The unfortunate phrase even found its way into movies like *The Little American*, a 1917 film starring Canada-born Mary Pickford, "the girl with the curls," which showed the kaiser dismissing the mere "scrap of paper."[4]

Germany's treatment of civilians raised the stakes. When the kaiser's troops crossed into Belgian territory on August 4, they found that Belgians were not easily pushed aside. Massive cannons forced the capitulation of key forts, but resistance remained, and German commanders suppressed it brutally. Whole villages were leveled to punish a few people. Critics of British wartime propaganda later asserted that it went too far in painting the Germans as "Huns," yet the kaiser himself first used the term to describe his troops during the earlier Boxer Rebellion in China. And there was no gainsaying the open massacre of 674 civilians in the tiny town of Dinant on August 23, 1914, or the deliberate torching of medieval Louvain two days later to make examples of anyone defying occupation.[5]

America's traditional neutrality heightened its sympathy for Belgium. Dating from George Washington's Farewell Address of 1796 and strengthened by the Monroe Doctrine of 1823, the "Great Rule" had long proscribed intervention in Europe's blood feuds. Yet despite this commonality with Belgium, or because of it, Woodrow Wilson pledged to stay on the sidelines.

"We must be impartial in thought as well as in action," the president told the public in August 1914. America must not be "thrown off our balance by a war with which we have nothing to do, whose causes cannot touch us," he lectured Congress.[6] But neutrality was hard to maintain as the war dragged on. At first, German victory seemed imminent. Gigantic artillery demolished Belgium's impregnable fortifications at Liège, and the kaiser's soldiers poured into France. Paris was roughly 250 miles away. A river of well-equipped troops advanced to within ten miles of the capital. Fighting was so close that French reinforcements took Parisian taxicabs to the battle on the Marne River.

Had the war ended then, in September 1914, the United States would never have been implicated. But French and British forces beat off the invaders, the Russian army tangled with German troops on the eastern front, and Serbians punched back in the Balkans. The Ottoman Empire entered the war. Fighting swept the Middle East. Troops in Belgium and France settled down into trenches on a front soon dubbed "the meat-grinder" for the way it chewed up men.

Germany introduced a score of new technologies with dire consequences for civilians, not considered legitimate targets

by the understood but unenforced laws of war. This, too,
pricked the consciences of Americans and other neutrals.
Leader of the world's chemical industry, Germany pioneered
the use of fatal chlorine gas against enemy forces near country
villages, spraying men like bugs. Beginning in early 1915,
German zeppelins and airplanes bombed Paris and London
from the air, another first in warfare. In Britain, they killed
366 women and 252 children. A soldier who survived the
perilous western front came home to find his wife and two
children dead from a German bomb in Leicestershire.[7] Several countries possessed submarines, a new naval technology, but Germany's had greater speed and firepower. More
important, the German government used them in ways un-
constrained by what its strategists dismissed as *Humanitäts-
düselei*, "humanitarian babbling."[8]

To the American public, the heinous use of scientific advances was epitomized by the May 1915 attack on the RMS
Lusitania, carrying 1,960 men, women, and children along
with munitions. Some of the nation's richest men were aboard,
including Alfred Vanderbilt, heir to the great Vanderbilt
fortune, along with other prominent industrialists, enter-
tainers, and socialites who could afford first class. One Philadelphia family was traveling with six young children.

Everyone felt the tremendous shudder as the ship took the
hammer of a single torpedo, which triggered a second explo-
sion inside the bowels of the luxury liner. Eleven miles from
Ireland's green hills, the *Lusitania* listed to starboard. Within
eighteen minutes, the huge ship nosed downward, exposing
its immense bronze propellers in the rear, and plunged to the
bottom. Only six of forty-eight lifeboats were launched in
time. Of the 1,191 killed, 125 were U.S. citizens.

Americans saw these and other tragedies splayed across the front pages of small-town newspapers, which readers combed closely in an era before electronic distractions. They cheered instinctively for the underdog. Before the war, many had admired Germany, often assumed to be the most advanced, innovative European nation. But in 1914, Berlin was not threatened. Paris was.

Fearing the city's imminent fall, the French government moved its operations to Bordeaux in September 1914. To transfixed onlookers, *Paris* stopped being a synonym for government, like *Washington,* and became merely the storied city of lights. Willa Cather, the great novelist of the American heartland (and a former math student of John Pershing, when he was stationed in Nebraska and taught prep school on the side), won the Pulitzer Prize in 1923 for her depiction of the mood change that came over many U.S. households.[9]

Cather's young Nebraskan protagonist in *One of Ours* smiled at his mother. She had "always thought of Paris as the wickedest of cities, the capital of a frivolous, wine-drinking, Catholic people who were responsible for the massacre of St. Bartholomew and the grinning atheist, Voltaire." But ever since the French army began falling back, from Lorraine in the east and Amiens in the north, "he had noticed with amusement her growing solicitude for France. . . . Paris seemed suddenly to have become the capital, not of France, but of the world!" The protagonist, like others, hung on the news during sieges like the Battle of the Marne outside Paris: "For four days men watched that name as they might stand out at night to watch a comet, or see a star fall."[10]

The European tragedy trickled into nickelodeons. *The Little American* showed Mary Pickford surviving the German torpedoing of the *"Veritana,"* which rhymed with *Lusitania* and reminded moviegoers of victims who did not make it to lifeboats. A 1916 Cecil B. DeMille extravaganza on Joan of Arc pricked memories of ancient Lorraine. A New York advertisement for *Joan the Woman* spelled out the connection: "Soldiers in the trenches today, fighting on French soil, to preserve for themselves and the future what Joan of Arc gave them, declare her figure hovers over them on the field of battle." Government posters later featured the Maid of Lorraine to sell war bonds: "Joan of Arc Saved France; Women of America Save Your Country."[11]

The war was terrible and fascinating. It was not just the most momentous event of the time. It was one of the most momentous events of all time. Europe was engulfed end to end. Medieval towns were pulverized, cruise ships sunk within sight of land, a generation shattered. For anyone following events, it was hard not to care.

American dollars were also at stake. A key dimension of the nation's traditional neutrality was its insistence that nonbelligerents had the right to export food and other nonmilitary goods to countries at war. The oceans were commons. Countries not in the fight should be able to trade with whomever they wished without reprisal.

It was an ancient theory of respectable lineage that dated to the Peace of Utrecht between Spain, England, France, and Holland in 1713 and 1714. The United States had made profitable use of the principle as early as the Napoleonic era, selling to all sides in the wars that engulfed Europe, until British disregard for the enlightened notion dragged Washington into war in 1812 and America's own capital was burned to the ground.

A century later, the United States placed itself again in the same precarious position. It might have chosen to stay out, but selling merchandise to both sides was enticing. Like other neutrals, such as Denmark and Norway, America benefited from the bonanza of foreign war. In 1914, the United States got out from under debt for the first time in more than a century.

Britain and Germany declared blockades to deter further sales. Once the conflict on land settled into stalemate, starving one's enemy by water became paramount. Britain laced the North Sea with mines. Several American ships collided with them and sank. German submarines stalked the Atlantic and the English Channel, torpedoing neutrals. Norway grudgingly accepted the destruction of 436 ships as the price of doing business. The United States could have done the same, or American ships could have avoided the war zone. Congress debated restrictions on travel but decided to honor principle and profit above prudence.[12]

To many, German U-boats seemed especially devilish compared with blind, inanimate mines. Fully informed captains looked right through their periscopes at passenger ships. The wily predators were also extremely hard to catch. During one week of 1916, three U-boats in the English Channel evaded detection in an area "patrolled by forty-nine destroyers, forty-eight torpedo boats, seven Q-ships, and 468 auxiliaries—some 572 anti-submarine vessels in all, not counting aircraft," one report noted. Those three U-boats destroyed thirty surface ships before escaping the massive dragnet unscathed.[13]

Hundreds of thousands of tonnage sank each month, mostly Allied. According to a prominent American observer on the ground in London, the people of Britain, France, and Italy were down to three weeks' worth of wheat in 1917 due to ship losses. "The situation was dangerous almost beyond description," reported Herbert Hoover, then a private citizen.[14]

Germany gambled on keeping the United States out of the war. At Woodrow Wilson's request, Kaiser Wilhelm pledged to restrict attacks on neutral vessels after the loss of the *Lusitania* in 1915. American public opinion calmed. Wilson tried repeatedly to mediate an end to the conflict, and in January 1917 he encouraged all belligerents to accept a nonpunitive settlement of their differences, what he called a "peace without victory."

Some historians portray the United States as an empire eager to impose its will around the world in the twentieth century, yet Wilson's long delay in entering the war belies this interpretation. The president hovered nearly three years between the conviction that the European conflict could not touch the United States, as he said in 1914, and the tortured realization in 1917 that "neutrality is no longer feasible or desirable where the peace of the world is involved." He called the duty of military intervention "distressing," "tragical," and "oppressive." Wilson and other Americans were reluctant to become entangled. The president felt his way into a leadership role, observing protocols established by the Hague conferences of 1899 and 1907. Doing so, Wilson employed tools authorized by new treaties that encouraged "strangers to the dispute" to mediate conflicts before they reached the Permanent Court of Arbitration, a fledgling international body established to prevent war. The U.S. approach was consultative, not imperial. As Wilson told Congress, "We are but one of the champions of the rights of mankind."[15]

Germany had dragged its feet at the Hague conferences, however, spurning proposals for compulsory arbitration. Now the kaiser believed he was close to beating his enemies, so Wilson's proposals of mediation held little appeal. Wilhelm resumed his sink-on-sight policy on February 1, 1917. U-boats attacked unarmed U.S. merchantmen once more, sinking the *City of Memphis, Illinois,* and *Vigilancia* in two days. Hospital ships and vessels carrying food to Belgium went down as well.[16]

German diplomats meanwhile plotted a military alliance with Mexico should the United States retaliate by allying with Britain and France. Mexico would be rewarded with territory from Arizona, New Mexico, and Texas. The January 1917 publication of the secret Zimmermann Telegram from Berlin to the kaiser's embassy in Mexico City, outlining the plot, exposed German machinations to an initially incredulous, then angry, American public.

U.S. neutrality ended. On April 2, 1917, President Wilson asked Congress to "formally accept the status of belligerent which has thus been thrust upon it." Not everyone agreed that America had no other choice. Senator Robert La Follette of Wisconsin, a widely esteemed Progressive, cautioned against adopting European problems. British underwater mines had taken innocent lives as ruthlessly as German U-boats. America's poor would be "called upon to rot in the trenches" to quench industrialists' desire for markets. Average citizens opposed taking sides, La Follette claimed, reading telegrams and letters aloud to the Senate. In one, a North Dakota farmwoman complained, "The maudlin sympathy with women who lose their lives on the high seas these days is ridiculous. Let the women stay at home, where they belong. . . . I protest, in the name of humanity, against the taking away of husbands, fathers, sons, and brothers to be butchered."[17]

Yet the majority agreed with Wilson. Germany had driven U.S. merchantmen from the open seas. America must defend its rights. Senator Morris Sheppard of Texas quoted other women: "The Rebecca Stoddard Chapter, Daughters of the American Revolution, of El Paso, Tex., heartily endorse President Woodrow Wilson in his platform of preparedness and patriotism." Men and women sent letters from Cherokee County as well: "The citizens of Jacksonville, Tex., in mass meeting assembled adopted resolutions unanimously indorsing the action already taken by President Wilson . . . and will indorse any means he and the Congress may adopt in the future for the protection of American rights, lives, and property."[18]

Supportive senators piled on. "The hour is here," Democratic senator Henry Myers of Montana asserted. "No one denies that German submarines have unlawfully murdered our citizens, sunk our ships, destroyed our property. . . . What shall we do about it?" Senator Claude Swanson of Virginia pointed out that Germany had treated its promise to stop attacking neutral ships as yet more "'scraps of paper,' to be utterly disregarded."[19]

Arizona senator Henry Fountain Ashurst concurred. Possessed of such a keen sense of drama that *Time* magazine later called him "the longest U.S. theatrical engagement on record," Ashurst argued, "Democracy will not survive if in times of danger it does no more than preach the doctrine of philosophical nonresistance, [or] simper sentimental regret over a deadly wound it receives." A supporter of women's suffrage and other Progressive causes, Ashurst gave a speech overflowing with lofty resolve. The United States should not only "build cities, conquer deserts," but also "heal sore wounds, crush bigotry and race hatred, struggle for liberty."[20]

Both pragmatism and grandiosity were old strains in the national personality. Of the 695 neutral ships torpedoed and sunk worldwide between 1914 and 1917, only 20 (less than 3 percent) were U.S. merchant vessels. Yet a cut against honor was no less galling for being a nick. The world's first postcolonial nation viewed itself as a defender of larger ideals. Outrage at Germany's harms became intertwined with optimism that America might help defend free peoples worldwide.[21]

The Senate accepted the president's recommendation by a vote of 82 to 6, the House by 373 to 50. The first female ever to sit in Congress voted no to express her pacifism. Republican suffragist Jeannette Rankin represented Montana, one of eleven states where women could then vote.

Some were aghast at the declaration of war, but many felt about the defense of democracy the same way Willa Cather's Nebraska farm boy did about Paris: "There was nothing else he would so gladly be as an atom in that wall of flesh and blood that rose and melted and rose again before that city which had meant so much throughout the centuries—but had never meant so much before." One-third of the men who served in the U.S. armed forces volunteered. All of the women did. Indeed, voluntarism characterized the American system of government and American way of war.[22]

The United States entered the European brawl on April 6 with an unusual handicap. Not only was it militarily unprepared, but its authority was weak compared with that of its rivals. Washington was supposed to perform only narrow duties assigned it by the Constitution—primarily defense, rarely required given America's wide oceanic buffers.

Ostensibly, the forty-eight states reserved all undesignated powers. Outside military bases, the District of Columbia, and the territories, the federal government operated not a single system of education, sanitation, transportation, communications, police protection, or most anything else that central governments provided elsewhere in the industrialized world. Delivering the mail was its biggest activity. In contrast, countries like England, France, and Germany ran public utilities like telephones and sustained large, ongoing military establishments.[23]

Washington's limited scope fostered reliance on private groups for public tasks. The federal government repeatedly turned to the private sector in the early twentieth century to perform work for which it was not authorized, from building railroads and gasworks to caring for the poor, policing dissidents, and mobilizing for defense. A rich tradition of voluntary associations, noted as far back as Alexis de Tocqueville and by historians since, undergirded this policy. The famous Frenchman observed in 1835 that no other country in the world depended so thoroughly on "the agency of private individuals," harnessed in associations, "to promote public order, commerce, industry, morality, and religion." Preferring minimal government, which they equated with liberty, citizens generally did not expect Washington to do things for them. If anything, they felt obligated to do things for it. But the extraordinary emergency of a world war pushed public-private cooperation to levels that practically erased the distinction—and created long-term complications.[24]

A famous example was the Commission for Relief in Belgium, which an engineer living in London organized after the Great War broke out. The forty-year-old Herbert Hoover had worked around the world, from Australia to China to England. An Iowa orphan who got into a good college and became a self-made millionaire, Hoover excelled at bringing public and private groups together for ad hoc projects.

In 1914, the Stanford University graduate helped organize the evacuation of the 200,000 American citizens trapped in embattled Europe, an emergency service that embassies normally spearheaded. Afterward, Hoover raised millions of dollars and organized fifty thousand volunteers worldwide to succor Belgium, even negotiating with the German government to ensure deliveries of food to civilians. The *New York Times* reported in 1915 that municipal authorities of the shattered town of Louvain named streets for George Washington and Woodrow Wilson by way of thanks.[25]

In 1917, Wilson appointed Herbert Hoover head of the U.S. Food Administration to coax Americans into eating less in order to feed troops, allies, and hungry Belgians. Unlike in Britain and Germany, there would be no government rationing. Instead, every man, woman, and child would link hands with their government. As reported in *Bell Telephone News*, Hoover invited "all classes and all trades to sign a volunteer pledge to cooperate with us in the undertaking and so become as much members of the Food Administration as we ourselves are."[26]

The Red Cross, Salvation Army, and other ongoing groups also filled the gap between social need and government capacity. The army had no reserve hospital corps, for example. The Red Cross organized, trained, and paid special units at civilian hospitals in the United States from 1915 to 1917 to prepare for an emergency. Once America entered the war, the army inducted former Red Cross doctors and nurses and placed them under government authority. This wasn't surprising. Civilian aid in foreign crises had been the first resort of government ever since Civil War nurse Clara Barton started the U.S. branch of the International Red Cross in 1888. Members of the American Red Cross believed deeply that their nation had a benevolent, compassionate role to play in the world and that citizens should get busy. By war's end, roughly one-third of Americans, including eleven million children, had joined.[27]

Women's organizations jumped at the opportunity to contribute, too, hoping to prove their competence at new civic roles. Wellesley College purchased and transported an ambulance to France in memory of General Pershing's late wife, Frankie—a beloved graduate of the Wellesley class of 1903. Smith College, another of the elite "Seven Sisters" group of East Coast women's colleges, sent a relief unit of professors and graduates, who assisted in the French refugee crisis for more than a year. Anna Howard Shaw, who preceded Carrie Chapman Catt as NAWSA president, headed the Woman's Committee of the Council of National Defense, an umbrella group that brought businesses and private organizations together with government to coordinate mobilization.

Class background shaped voluntarism. Middle-class women did not need or expect to be paid, and officials were relieved not to fork over money. Clubwomen organized in extraordinary numbers to sell war bonds, serve sandwiches to soldiers at railroad stations, roll bandages, knit socks, and prepare "comfort packages" with toiletries and hygienic advice for boys overseas. In the city of Los Angeles alone, twenty-nine thousand women volunteered in eight hundred precincts organized by the Woman's Committee of the Council of National Defense.[28]

But of all World War I collaborations, few were as unprecedented as that between the profit-making telecommunications industry and the war-making Signal Corps. On the day Congress declared America's belligerency, legislators also entertained a bill to commandeer the vital telephone system of the District of Columbia to ensure "complete control of such means of communication in safeguarding its military and executive affairs." The United States eventually nationalized all telephones, telegraphs, and railroads. The farsighted cooperation of the private sector helped ensure this was only for the duration of the war.[29]

Telephones solved ancient problems of communication in battle. At the time of the American Revolution, silversmith Paul Revere posted lanterns in the tower of Old North Church in 1775 to warn of approaching British troops. The alarm succeeded only because Revere also organized thirty riders to deliver the message once they saw the lamps. Communication failures were equally legendary. Just seven years before the American Civil War, the 1854 Battle of Balaklava—immortalized in Tennyson's "Charge of the Light Brigade"—resulted from a messenger pointing straight ahead instead of sideways when delivering an order about which direction British troops should attack during the Crimean War.

The homely telephone changed all this, becoming one of the war's most important weapons overnight and turning women into soldiers. A chronicler of the American Expeditionary Forces called them the *Circuits of Victory*.[30] An early silent film dubbed them *The Whispering Wires*. The wires first connected by Alexander Graham Bell linked private interests to public ones and ran right into the muddy trenches of no man's land. In 1917, telephones became the central nervous system of the U.S. Army. Switchboards were its synapses.

For the first time, officers and troops positioned leagues apart communicated easily, once an operator connected them.

Telephones coupled the front lines with command posts out of range and supply depots hundreds of miles away. In World War I, when millions were deployed on a front too vast for commanders to see the full scope of any battle, and massive guns tossed the forests of France into the air like matchsticks, telephones allowed men under fire to tell unseeing officers what was happening. In the words of French novelist Sébastien Japrisot, the phone wire that "snaked its way through all the trenches, through all the winters," was soldiers' primary "link to the world of the living."[31]

Radios, a wireless form of sound transmission, did not yet carry voices, and Morse code signals reached only five to twenty-five miles on land. Field radios were so heavy in 1916 that three mules were needed to pull a single unit.[32] Additionally, anyone could intercept the frequency. Colonel Parker Hitt, head of communications at Pershing's French headquarters, called radio "a last resort" for any commander. "The enemy is sure to copy all radio messages sent out and at the same time will accurately locate the position of the sending station and usually tell what kind of a headquarters it is serving."[33]

In contrast, lightweight telephone wires linked front and rear with a private, spontaneous means of communication reaching hundreds of miles. Linemen could run wire on foot. Telephones facilitated troop movements, supply orders, and moment-by-moment tactical decisions. British historian John Terraine writes, "The chief instrument of generalship throughout the war was, of course, the telephone." Given the size and fury of the battlefields, "Only the telephone, the uncertain, temperamental telephone of the second decade, gave generals any real power of command."[34] Women, it so happened, were masters of this new technology—the only one in which Americans led during the war.

Invented in Massachusetts in 1876, the phone was the natural love of a chatty people scattered across thousands of miles in a rugged land. One year after Alexander Graham Bell's first public demonstration in Salem, Massachusetts, a Connecticut druggist established a telephone network connecting his store with a horse-and-buggy delivery service. A self-contained system like this, linking a single business internally or with its clients, was a "private branch exchange," or PBX. The following year, forty miles farther south, twenty-one unrelated individuals subscribed to the first public system in the world.[35] Within a few years, Bell's company, incorporated as American Telephone and Telegraph (AT&T), fashioned a "Grand System" of "exchanges" that linked thousands and eventually millions of people. Connections through larger and larger switchboards transferred one so-called subscriber to another.[36]

Although telephones were designed to be easy enough for children to use, Bell employed a burgeoning army of specialists to keep the network humming: splicers, installers, linemen, and engineers. Most prominent of all were operators. For the first fifty years of the Bell System, every caller—from John Doe to John Pershing—spoke first with an operator, whose simple job was extraordinarily complicated.

Telephones then had no dials. The subscriber lifted her or his receiver from a black phone shaped like a tall candlestick. Day or night, this tripped an electrical charge at a distant switchboard. The current lit a tiny bulb or caused a small metal flap to "drop" above the associated line. (Today's term for a dropped call echoes this early terminology.)

Subscribers did not need electricity in their homes—a feature that soon proved invaluable on battlefields. "Common" batteries housed in central telephone exchanges sent power down the lines to the subscriber. In rural areas out of range of a common battery, a subscriber cranked a handle to generate the impulse that tripped the light or flap on the operator's board in town. These hand-cranked instruments were called magneto phones for the internal magnets that created a charge.[37]

Once alerted, a headset-wearing operator reached for a flexible cord with male plugs on both ends. The operator inserted one of the plugs into the female jack for that particular subscriber, located in a row at the bottom of the switchboard.

"Number, please?" she would ask.

The subscriber replied through a conical funnel on the top of the candlestick, and the operator sprang into action. The top of her board held rows upon rows of numbered jacks that corresponded to all the subscribers in the exchange. Other operators connected with them by reaching across one another. If a caller wanted someone in the operator's immediate network, she merely inserted the opposite plug into the relevant jack. Then she turned a key or pulled a peg to ring a bell that notified the person on the other end.

Most early numbers were party lines. Operators devised specific patterns to alert up to sixteen subscribers whether they or someone else down the line should pick up the receiver. Rings could be short or long, and the operator might buzz two, three, or four times. Customers sometimes gave standing instructions to ring quietly when babies were sleeping or as loudly as possible if there was competing noise, as in a factory. Operators were expected to remember individual combinations and preferences.[38] After the operator made the connection, she checked back periodically to see if the parties were still talking or the cords should be unplugged.[39]

Long-distance connections were more complex. If a subscriber wished to call outside the network, a local operator first told the subscriber to hang up. She then rang a long-distance operator through a designated trunk line to a toll office in a bigger city. The first operator told the toll operator which number to call. If the desired person was several cities away, the first operator stayed on the line until a relay was built up—operator by operator, trunk line by trunk line. She then rang the original subscriber to report that the desired party was on the line. If the distance was great, it might take a local operator working with the toll office several hours to complete the connection—all the while taking new requests, ringing idiosyncratic subscribers, counting local calls for billing purposes, and timing long-distance tolls.

As of 1900, 80 percent of operators were female. Young men dominated telegraph offices, where Morse code messages were translated one at a time and required no personal interaction with customers. But the telephone systems that sprang up across the country preferred women. Gender expectations trained females from birth in the genteel diplomacy required for conversing with impatient callers. Women also showed great nimbleness with the multiple jacks, sockets, ringers, and buzzers on the boards of busy switching stations. It was a fast-paced job requiring steady nerves.

Scientific studies a century later suggest that females may possess advantages over males in multitasking. The U.S. Census Bureau foreshadowed this notion in 1902: "It has been demonstrated beyond all doubt that the work of operating is better handled by women than by men or boys." As a consequence, the bureau reported, "The telephone companies in the United States have been alive to the importance of securing and retaining this quality of labor."[40]

American companies also used women to promote business. Telephone service was a luxury. Personable customer service by a courteous (and possibly beauteous) unseen operator was part of its appeal. Supervisors expected exquisitely good manners, clear speech, excellent hearing, impeccable performance, and crisp efficiency. Immigrants with accents were excluded, as were African Americans. Union organizers intent on protecting jobs echoed AT&T's idealization of operators as elegant but democratic public servants ready to serve the humblest customer.[41]

This business strategy may account for a strange technological delay in the United States, where operators connected calls long after companies elsewhere had mechanized. England, Sweden, and Belgium adopted automatic switching well before AT&T, even though a Kansas City undertaker—disgusted with the favoritism shown his competitor by an operator at the phone company—invented what he called the first "girl-less, cuss-less, out-of-orderless, and wait-less telephone" in 1891.[42] Despite these American origins, AT&T did not adopt mechanical switching for local calls until after the Great War. Long-distance calls required operator assistance into the 1970s.[43]

Until then, women were the voice of AT&T. Diligent and quick, the best of them efficiently manipulated cords and plugs while fielding requests for the time and other information. One operator in Montana observed, "On a busy day these cords were woven across the board in a constantly changing, confusing pattern." Yet the job was also oddly satisfying. It gave operators "an awful responsibility toward our little corner of the world," wrote Dorothy Johnson. "We really helped keep it running, one girl at a time all by herself at the board." The most accomplished could "ring a number (front key plus a button for L, K, Y, or X) with the left hand while flipping a back plug into a hole with the right hand and caroling 'That party doesn't answer.' . . . Those girls' hands darted around like a pair of hummingbirds."[44] In urban exchanges, operators completed hundreds of calls per hour, scanning thousands of tiny illuminated jacks and sweeping the board of old connections.[45]

Women had few other opportunities for well-paid, white-collar employment. Even some college graduates aspired to telephone work. A spinner in a clothing factory might earn $1 per day, while an operator earned $2.50. Telephone operating paid better than most female occupations, with the exception of teaching. "These jobs were much in demand," one operator later recalled. There were always hopefuls in the wings waiting for the enviably employed to "get married, move away, or drop dead."[46]

By World War I, telephone wires branched like fine capillaries over cities and towns, going right into homes. In the West, telephone lines were strung through bushes or tied to trees. Barbed-wire fences substituted nicely if one possessed a magneto phone.[47] Telephones proved especially useful in rugged terrain, another characteristic essential for military purposes. In national forests, the first park rangers carried three-pound phones. Whenever he needed to reach civilization, a man simply threw a wire across any telephone line already running through the forest and attached both ends to his own phone. Isolated farmers and residents of the far West became the telephone's most ardent users. San Francisco had the highest telephone use of any city in the world, with Los Angeles close behind.[48] The United States owned 70 percent of the world's telephones, though its population was roughly half that of Russia and not much bigger than Germany's. At only 9 percent, Germany had the second highest teledensity. France was far behind, with 2 percent.[49]

America's gigantic distances propelled further innovation. To connect East to West for the 1914 opening of the Panama Canal, AT&T began work on a device called a repeater. The repeater replaced "loading coils"—amplifiers previously mounted on outdoor poles every eight miles to boost sound waves that faded, blurred, and acquired so much static with distance that they became nearly unintelligible. Using high-vacuum tubes, repeaters were more delicate and expensive than loading coils but could function 150 miles apart.

In the early twentieth century, telephones in the United States reached farther, conveyed more messages on their bundled circuits, and reproduced sound with greater fidelity than anywhere else in the world. When the British commanding officer in World War I used an American-built line from France to England, he exclaimed, "Would you believe it? They actually recognized my voice in London before I told them who I was!" In fact, telephones were the only military technology in which the United States enjoyed clear superiority. The United States had no manufacturing facilities for large artillery. As General John Pershing acknowledged, the American Expeditionary Forces relied almost entirely on French cannons, tanks, and planes.[50]

Bell Telephone finally connected all of North America on July 29, 1914. Calling San Francisco from his New York office, white-haired AT&T president Theodore Vail conversed quietly with the president of Pacific Telephone and Telegraph. Publicists trumpeted the event with an advertisement portraying science as a telephone goddess holding a candlestick phone aloft tiny New York and San Francisco, one wire connecting them. The operator embodied progress much like the figure of Columbia leading the way across the prairie in the nineteenth century.[51]

On the very same day, Austro-Hungarian warships shelled Serbia from the Danube River to punish it for spawning terrorists like Gavrilo Princip, the nervous nineteen-year-old who had assassinated Archduke Franz Ferdinand a month earlier. Allies of both parties looked on, trying to decide whether to step in.[52]

The human voice had traveled three thousand miles. The human race was poised on the brink of the inferno.

Merle Egan helped relay the first transcontinental calls that took place while war blazed in Europe. "I was fortunate enough to 'listen in' on the first conversation from Washington D.C. to the West Coast," she told a later AT&T chairman.[53] Merle then worked for Mountain States Telephone and Telegraph. Born on a Kansas farm, she grew up in the coal-mining camps of southern Colorado, where her father was superintendent of mines until fired for his rabble-rousing and fiery Irish oratory.

At the time, fewer than 15 percent of young Americans graduated high school, and Merle was no exception. She completed three years before economic circumstances sent her to work, first as a relief operator, then as a toll operator in the PBX exchange of Denver's proudest hotel, the ornate Brown Palace. "From that July day in 1906 when I became an operator in Denver I was fascinated by the fact that people could communicate thru the cords I handled," Merle wrote. It was like holding lives in her palm.[54]

When opportunity arose in Montana to join a public system, Merle jumped at the chance. She proved so resourceful that the company sent her from "town to town to straighten out trouble." By the time World War I began, Egan was a "traffic supervisor" with eleven years' experience. She was also twenty-nine, unmarried, and being courted at a slow pace by a man in no hurry to make plans. Merle Egan was a fit, well-proportioned 165 pounds, taller and bigger than many men. It must have been hard to be five feet eleven in an era when the average male topped out at five feet six. One 1918 photograph shows a dark-eyed young woman with a wary smile, tilted chin, and skeptical, intelligent expression.[55]

Her employer described her as "industrious, level-headed, even tempered and cautious"—with instinctive tact in difficult situations. "Her individuality has made her a natural leader," he reported to the U.S. Army. "Her loyalty is unquestioned."[56]

Merle Egan was just the sort of person to show her grit in a patriotic cause—or any other campaign about which she cared. And she wasn't alone. Thousands of women felt exactly the same.

3

LOOKING FOR SOLDIERS AND
FINDING WOMEN

O N THE DAY Congress declared war, the Washington
office of the Signal Corps had exactly eleven officers
and ten men. Additional troops scattered around the country
brought the unit total to 55 officers and 1,570 enlisted men,
most maintaining telegraph cables to Alaska or servicing PBX
systems. Almost overnight, their job became to provide communications for an army of four million, nearly half of whom
would be stationed in a foreign country four thousand miles
away, across an ocean. In Europe, approximately 1.4 million
Americans eventually served at the front. The majority of
soldiers supported them.[1]

To meet their responsibilities, the chief signal officer reported to Congress, "98 per cent of the personnel of the
Signal Corps must have some knowledge of telephone, telegraph, or radio engineering." The corps needed an additional
fifty thousand trained men to build and operate—sometimes
under bombardment—a communications system more than
four hundred miles long, with branches throughout France.
Where would the army find them? As with other tasks, there
was only one possible answer in the early twentieth century:
the private sector.[2]

Fortunately, Black Jack Pershing had made the acquaintance of AT&T's chief engineer, John Carty, a year earlier. The occasion was a surprise phone call to Pershing, then camped with his forces along the Mexican border in El Paso, Texas. The National Geographic Society was holding its annual dinner in Washington, D.C., and asked Carty to demonstrate AT&T's new long-distance capabilities to eight hundred guests gathered for the glittery event. Each tuxedoed or bejeweled diner was handed a wired candlestick receiver to hold.[3]

Operators from Washington to El Paso relayed the unprecedented call nearly two thousand miles. Pershing's adjutant summoned him to the phone. With the secretary of war seated next to Carty at the head table, and the secretary of the navy in the audience, the AT&T engineer informed Pershing that hundreds were listening.

"If I had known it," the surprised general replied, "I might have thought of something worthwhile to say." A biographer wryly observed of the stoic, square-jawed Pershing that he "was a soldier and soldiers do not talk." However, like most senior officers Pershing was also one-quarter politician. He warmly complimented the National Geographic Society for its "splendid work."[4]

Over the next year, John Carty kept in contact with the American military. In June 1916, during the nine-month Battle of Verdun that slew more than 300,000, President Wilson signed the National Defense Act, also known as the Hay-Chamberlain Bill, to establish an Officers' Reserve Corps to induct civilians with specialized skills. Six months before the United States entered the Great War, John Carty donned the uniform of a Signal Corps major. Top officials of Western Electric Company and Western Union Telegraph did, too.[5]

In January 1917, the chief signal officer of the U.S. Army, George Squier, traveled north to New York City to visit John Carty and AT&T president Theodore Vail. The trio met at AT&T's new headquarters at 195 Broadway, over which towered "Golden Boy," a twenty-eight-foot gilded statue of a winged male figure clutching bolts of lightning. Officially titled *The Spirit of Communication*, Golden Boy was the largest statue in New York after the Statue of Liberty.[6]

A West Point polymath with a piercing expression and a PhD in physics from Johns Hopkins, General George Squier was well equipped to appreciate the complex tasks ahead. At the time, the army assigned anything new and experimental to the Signal Corps, then the army's scientific branch.[7] In 1907, Squier organized the U.S. Army Aeronautical Division as a reconnaissance wing of the Signal Corps. The first military passenger in a Wright brothers' prototype, Squier purchased the handful of airplanes that became the U.S. Air Force. After the war, the inventive veteran who loved gadgets started a company called Muzak to pipe music over electrical wires.

In his January 1917 meeting with Carty and Vail, George Squier proposed mobilizing AT&T. As he put it, "officers in the uniform of their country" would serve the U.S. Signal Corps from their regular corporate offices and take orders from the government while running private enterprises. In army dress, businessmen would recruit battalions from within their organizations, harnessing "the ablest and most experienced technical men in the country in the shortest possible time."[8]

AT&T had patriotic and economic incentives to respond favorably. When Wilson first took office, peacetime advisers urged the Progressive president to expropriate telecommunications altogether. In most other advanced industrialized

countries, governments operated phones as a public utility
like water or gas. Theodore Vail invested significant resources
into convincing the president that private companies operated more efficiently than the government could. In 1914,
AT&T published a three-volume tome incongruously titled
Brief Arguments against Public Ownership.[9]

On the same February morning that Wilson severed dip-
lomatic relations with Germany over unrestricted submarine
warfare, Major John Carty booked a train to Washington. The
president had not yet asked Congress for a declaration of war,
but as far as Carty was concerned, "the fight was on. From
that time forward, we were at war with Germany—at least in-
sofar as my own plans were concerned."[10]

Carty and other reservists sat down with government officials to prepare. They needed to double the number of phone
wires in and out of Washington; triple the number of new
transcontinental lines between Chicago and San Francisco;
and fortify the Coast Guard telephone system on the Atlantic,
Gulf of Mexico, and Pacific coasts. All of this had to be done
as if lives depended on it, which they would.[11]

Squier and Carty devised ambitious personnel plans as
well. AT&T would use its vast industrial network to locate
men for what became known as the Bell Battalions. When the
war finally came on April 6, 1917, thousands volunteered on
cue for twelve Telegraph Battalions and two Radio Companies, staffed entirely by AT&T employees in Signal Corps uniform. The battalions possessed 202 men each, plus officers
who had previously been their civilian supervisors.

In the words of one veteran, the Bell Battalions were composed of "cable splicers, linemen, switchboard men, test men,
helpers, truck drivers, motorcycle drivers, mechanics—each
man trained for years in his particular job, and each ready to

bet his hat that there was no other man in the world that could beat him at his own game." *Bell Telephone News* explained to employees in 1917 that "complete field signal battalions, officered by their own managers . . . were taken bodily into the service."[12]

Remarkably, AT&T, Western Electric, and Western Union all made it easier for patriotic employees to volunteer by ensuring they would not lose money doing so. The companies paid the difference between higher civilian wages and low army salaries for the entire war. If a man died, federal war risk insurance provided for his family. Some would, of course; running telephone lines into trenches or alongside advancing armies guaranteed it.

The French already knew this. One officer at the 1916 Battle of Verdun described the procedure for fixing broken telephone wires when enemy projectiles snapped the connection between his observation post and the artillery batteries behind him:

A man gets out at once for repairs, crawling along his stomach through all this place of bursting mines and shells. It seems quite impossible that he should escape the rain of shells, which exceeds anything imaginable. Our man seems to be enveloped in explosions, and shelters himself from time to time in the shell craters which honeycomb the ground. Finally, he reaches a less stormy spot, mends his wires, and then, as it would be madness to try to return, settles down in a big crater and waits for the storm to pass.[13]

In April 1917, however, the American military had yet to see this type of action. Industry and government were still exploring the technical requirements of an industrialized war. A few days before the first convoy sailed for France with General John Pershing and the initial troops of the American Expeditionary Forces (AEF), Major Carty met the army officer who would command the Bell Battalions overseas.

George Squier had chosen Brigadier General Edgar Russel for the job of "Chief Signal Officer, AEF." A southern, well-mannered career officer with movie star looks, Edgar Russel met with Carty at AT&T's New York headquarters. There they debated which type of equipment to send to the front. Should the army rely on older equipment that could take battlefield punishment, or newer, faster, but more fragile devices? Should it employ sturdy loading coils stationed every eight miles, or fragile vacuum-tube repeaters every hundred miles? Could it get away with newer lightweight circuits that allowed six telegraph messages and two phone calls to pass over the same line, or would these break more easily than heavy, old-fashioned wires that transmitted fewer messages?[14]

Major John Carty gave his opinion: use the best technology.

Brigadier General Edgar Russel agreed. He pressed Carty for whatever AT&T could spare. When Pershing and Russel sailed from New York shortly thereafter on the converted English luxury liner *Baltic*, they made room at the last minute for two telephone engineers. Stowed in the hold were Western Electric PBX switchboards diverted from a rush order expected in Chicago and a few dozen telephone poles borrowed off a Pennsylvania railway siding.[15]

Merle Egan found similar switchboards the next year when she arrived in Tours, France, delighted to discover that General Russel had snagged "the same type of equipment used in the large telephone offices at home."[16] By then, the army had decided it must also have the best personnel to go along with it.

The army had not planned to induct women. In fact, it was far behind other armed services in this regard. The U.S. Navy had steamed ahead, abreast of the British War Department.

Anticipating the war, Secretary of the Navy Josephus Daniels realized early in 1917 that he already did not have enough sailors for ships being built under the Naval Act that President Wilson had nursed through Congress when trying to defend neutrality. If fighting commenced, the shortage would worsen. One solution was to find a new source of labor for clerical tasks so that more men could go to sea. Since ships defined navies, Daniels may have thought females could fill landlubber jobs without wounding the navy's pride.

Secretary Daniels asked his legal counsel: "Is there any regulation which specifies that a Navy yeoman be a man?" When advised that no statute prohibited females, Daniels sent out the order: "Enroll women in the naval service as yeomen and we will have the best clerical assistance the country can provide."[17] (More than a third of the nation's office personnel were then female.) Despite some skepticism, the navy obeyed its civilian secretary. On March 17, 1917, the United States became the world's first modern nation to enlist females.

Before he joined Wilson's cabinet, the genial Josephus Daniels was best known as a North Carolina newspaper editor. Like many Progressives in the South, Daniels walked a fine line between reform and reaction. His father had been killed for ambiguous loyalties during the Civil War, and

Daniels grew up with a bone-deep understanding of the risks of nonconformity. An amalgam of contradictory impulses, he led a campaign for white supremacy at the turn of the century that disenfranchised blacks in North Carolina; two decades later, he opposed the Ku Klux Klan as overbearing and violent. But on women's rights (or at least the rights of white women), Daniels was more consistent. His widowed mother had raised him. As a newspaperman, he had publicly supported female suffrage for several decades and married an independent, high-spirited woman who supported suffrage, too. The secretary of the navy was determined to treat his new recruits with respect.[18]

In the month before war broke out, two hundred joined up. They were ranked yeoman (F), for female. When newspapers dubbed them "yeomanettes," Daniels objected: "If a woman does a job, she ought to have the name of the job." The secretary insisted on equal compensation, too. Female yeomen earned $28.75 per month, the same as men. The navy provided no housing for females, so it paid an additional $1.25 per day for billeting. Female recruits received free uniforms and medical care, and were eligible for war risk insurance. Most worked ten hours per day, six days a week, including night shifts. They wore the same insignia as men.[19]

Volunteers had to prove little beyond their age (eighteen to thirty-five) and willingness to serve. Some started the very day they walked into a recruiting office. When one young woman called her mother in Richmond with the startling news, her mother "was stunned into silence for a moment, then asked weakly, 'Oh, Sister, can you ever get out?'"[20]

Parents, the navy, and the general public had mixed reactions, but opposition soon turned into unaccustomed pride. The women's pluck, hard work, and stateside assignments mitigated criticism. So did government propaganda. In World War I, many governments communicated directly with their peoples through colorful, artistic posters that allowed citizens to visualize faraway problems in an age before television. Some showed pitiful females in need of rescue or aproned mothers skimping on bread to save food for soldiers. But another genre emerged, too: confident uniformed women embracing military service.

One of the navy's most memorable recruitment posters featured a short-haired, flirty woman in a sailor suit, saying, "Gee!! I wish I were *a man*. I'd join the Navy." Some versions of the poster included the tag line, "Be a man and do it."[21] But the poster went beyond shaming male "slackers." Depicting uniformed women as especially attractive and patriotic, it undoubtedly encouraged some females to enlist.

The navy began featuring women in recruitment drives and war bond rallies. Rear Admiral Robert Coontz, head of the Puget Sound Naval Yard, told a supervisor of the female yeomen under his command, "I want people to see we have girls in the Navy." Previously, only suffragists and temperance activists paraded en masse. Now female yeomen matched their strides to men, establishing new precedents for acceptable public behavior. After one group marched behind horses down Broad Street in Philadelphia, past Independence Hall, their drill instructor gave them an explicit lesson in the gender-neutral conduct expected of sailors confronted with manure. "You don't kick it, you don't jump over it, you step in it."[22]

Such training emboldened some young women. Like men, they yearned to serve "over there." One Chicago clerk stopped typing, fingers poised over the leggy keys, to tell her supervisor that female yeomen wanted to go to France. The grizzled navy captain looked at her, frowned, and said, "What the hell could a girl do on a battleship? Get back to your job."[23]

The nickname "yeomanette" persisted despite Daniels's protestations but may have helped some recruits reconcile their thirst for action with the uncomfortable, unfeminine implications. To the public, "yeomanette" suggested that military women were still "girls" and no one need get too alarmed. They enlisted for four-year stints with the guarantee that if the war ended sooner, they would be discharged with reserve status.[24]

Five months later, Josephus Daniels approved induction of women in the Marine Corps, then under the Navy Department. Wearing insignia identical to those of men, the so-called marinettes entered service in August 1918. They and the naval recruits were the first women admitted to full military rank by the United States. Nearly thirteen thousand joined.

The Marine Corps recruitment poster was bolder yet, showing men massed at the front and a self-confident, uniformed woman towering above them with one hand on her marine sword: "If You Want to Fight! Join the Marines." One of the first ten volunteers was the single mother of Ginger Rogers, later famous as an actress and a dancer. Sergeant Lela Rogers wrote articles for the Marine Corps newspaper *Leatherneck.*[25]

Josephus Daniels's attitude and the spirit of his volunteers may seem extraordinary given the context, yet they were emblematic of the changing times. The war created an entirely new role for both the United States and its citizens. Americans were unused to exercising responsibility outside their borders, or middle-class women outside their homes, but events now demanded it. Citizens' expectations were in flux. Elsewhere in the world, women were already filling unusual roles.

This was especially true of Britain, with its similar history of representative democracy and a strong suffrage movement. There the war bit deeply into population reserves. As men moved to the front, two million women replaced them in factories and offices. When evaluating the need for a draft in 1915, Parliament took the unprecedented step of requiring both sexes to register their occupations. Universal conscription of males began the next year, for the first time in Britain's ancient history. Pushed to the breaking point, the beleaguered War Office began eyeing women in 1917.[26]

Many suffragists supported the notion that women should help. Equal rights meant equal responsibilities. Emmeline Pankhurst called for obligatory national war service. Suddenly, her unorthodox ideas no longer sounded quite so preposterous. A renowned public speaker, the world's most infamous suffragist told followers the vote was meaningless if Britain was beaten: "We are fighting for our existence as a nation and all the ideals for which our forefathers have fought and sacrificed in the past." Emmeline's daughter, Christabel Pankhurst, criticized the kaiser's glorification of military might: "German *Kultur* means the supremacy of the male carried to the point of obscenity."[27]

British women turned the European crisis into an opportunity to demonstrate their worth as citizens. Two weeks after the war broke out in 1914, the major British suffrage organizations suspended militant action for the duration. In return, the government released all suffragettes imprisoned for civil disobedience. In British public opinion, women's suffrage came in out of the cold. This contrasted dramatically with the decade before, when His Majesty's government had repeatedly jailed Pankhurst, her two daughters, and hundreds of others for crimes that ranged from smashing windows and setting fires to obstructing Parliament. "I am what you call a hooligan," the dainty, impeccably dressed Emmeline Pankhurst told an amused Carnegie Hall audience in 1909.[28]

Now the British home secretary praised the Pankhurst matriarch as a patriot. The Ministry of Munitions even contributed £3000 to her recruitment efforts in 1914 on behalf of the British Army. For the first time, conservative spectators cheered rather than booed parades led by the Pankhursts.[29]

The inconceivable vote became a realistic possibility. In late 1917, the British Parliament debated enfranchising male soldiers excluded by property qualifications. The last legal barriers of class tottered under the war's weight. "Tommy," the nickname for the common British infantryman, had finally earned a right to a say in matters of war and peace. Gender restrictions suddenly seemed less fair as well. When Viscount Peel introduced the Bill for the Representation of the People in the House of Lords toward the end of 1917, he explained that "many have been converted by services rendered by women during the war." Other members concurred. As one argued, not only had women worked and suffered, but some had "died for their country." Let there be no mistake, he said, without women's "heroism, self-denial, skill and physical strength and endurance, this country would never have successfully faced the crisis."[30]

What were these services? Who had died?

Women's groups certainly helped with recruiting, and Red Cross ladies knitted, sewed, and rolled bandages. Mothers, sisters, and sweethearts kept farms going to feed the country and its troops. Working women stepped into men's shoes in war munitions plants. Some died in plant explosions, such as at No. 6 Shell Filling Plant in July 1918, when 139 civilians perished. Elsewhere, women handling explosives turned so yellow from exposure to toxic chemicals that they were called "canary girls." These women held British society together in its travail and saved His Majesty's government. Suffragists pointed to them as examples of civic solidarity.

But the impulse that expressed common purpose more dramatically—and rattled the gilded cage of Edwardian gender relations more alarmingly—than anything else was women's willingness to shoulder the military burdens of the Tommies fighting in France. Britain's uniformed women overseas provided the most conspicuous example of the kind of service that impressed Parliament, even while triggering ambivalence and prompting efforts to stem change. Desperate for personnel, Britain began recruiting women around the same time as the U.S. Navy. However, its terms were less egalitarian. Officials in the War Office took care to avoid any implication that women were "real" soldiers.

The British government recruited the first members of the Women's Army Auxiliary Corps (WAAC) in March 1917 and established the service officially in July. WAACs worked in France as cooks, mechanics, clerks, dockworkers, and switchboard operators to free males for the British Expeditionary Force (BEF). They retained civilian status to protect men's morale—and women's morality. They were paid less on the pretext that they accomplished less. The orders establishing the women's corps specified that four female clerks equaled three male soldiers. The insignia on their khaki uniforms depicted roses and fleur-de-lis rather than army crowns, crosses, and bars. Press releases emphasized the women's femininity. Privately, some officials questioned whether those who volunteered could possibly be the most moral girls. The War Office forbade WAACs from saluting or being saluted.[31]

British posters featured apple-cheeked women beckoning wholesomely to female viewers to join the WAACs and become "the girl behind the man behind the gun." German bombs killed nine WAACs a year later in France. The women were buried with full military honors. The *Times* of London wrote that they had "confirmed their right to khaki."[32] Queen Mary adopted the unit, thereafter called Queen Mary's Army Auxiliary Corps. Treated with a mixture of respect, paternalism, suspicion, and gratitude, more than eighty thousand served for the duration. They were a highly visible fraction of Britain's Army of five million.

Newspapers across the United States eagerly reported on the patriotic women who inspired the British Army to modify its "masculine traditions." Before the war, "the only military sphere for which women were thought eligible was nursing the sick and wounded," one Georgia newspaper told readers. Now, under the supervision of their own noncommissioned female officers, women drove ambulances, ran printing presses, and dug graves.[33]

Imperial Germany did not mobilize women, though a comparatively small percentage entered the domestic workforce for the first time, and some enlisted in a Home Guard to relieve men from police duties.[34] Russia did not recruit women until Tsar Nicholas II was deposed by a democratic revolution in February 1917. In the short period between this event and the Bolshevik coup d'état in October 1917, thousands of women volunteered. Some were college students; others factory workers and farmers' daughters. Approximately four thousand fought in combat.

Around St. Petersburg and Moscow, uniformed women "became a matter of course," according to pioneering California journalist Bessie Beatty, who chronicled the revolution in her 1918 book, *Red Heart of Russia*. They were "the most amazing single phenomenon of the war," she observed. "Not the isolated individual woman who has buckled on a sword and shouldered a gun throughout the pages of history," Beatty wrote, "but the woman soldier banded and fighting en masse—machine gun companies of her, battalions of her, scouting parties of her, whole regiments of her."[35]

The Russian soldiers carried carbines with bayonets. Casualties were heavy, and only 25 percent of "these modern Joans of Arc" survived, though they captured 102 German prisoners after one particularly fierce engagement. Just before the female battalions were disbanded, following the Bolsheviks' decision to sign a separate peace treaty exiting the war, AT&T regaled its women employees with the tale of their heroism. "We cannot but honor such women, or read of their doings without a thrill," *Bell Telephone News* reported in fall 1917. Bell reassured readers, however, that America needed female military personnel only for home duty in the U.S. Navy and Marine Corps.[36]

This was about to change.

Swelling training camps on American soil were the first to requisition women operators for the army. Under pressure to ship two million doughboys to France as rapidly as possible, cantonments like Fort Benjamin Harrison in Indiana and Camp Dix in New Jersey urgently requested help.

In September 1917, various camp commanders asked the army to build quarters for female switchboard operators. A colonel in the Quartermaster Corps wrote his superiors that Signal Corps officials believed "men operators cannot be obtained." Colonel I. W. Littell did not like the idea. Housing women alongside men went well beyond the navy's innovations. "In the opinion of this Division the employment of female telephone operators and particularly their housing on the reservation, is peculiarly undesirable and unwise," Littell argued. "To place a small number of young women in the midst of a large population otherwise entirely of men will inevitably lead to complications that will produce a flood of adverse criticism."[37]

Secretary of War Newton Baker shared this opinion. He did not even want the army to provide toilets for women. Three weeks after Littell's letter, on October 8, the War Department disapproved all requests to accommodate females. Division commanders should not "employ women in any capacity in the National Army cantonment and National Guard camps with the exception of female nurses attached to hospitals."[38]

Despite Baker's resistance, demand for switchboard expertise was insatiable. The army hired more and more civilian operators to manage PBXs. As a staff officer writing from Fort Monroe, Virginia, explained after the war, "The traffic is so heavy that only highly trained and experienced operators can handle it satisfactorily." Men's service assignments changed too frequently for them to develop expertise, plus "female operators are inherently more adept at this occupation than males [even] if the latter could be continuously retained on duty."[39]

When AT&T and the Signal Corps pointed out that virtually every army camp had no choice but to recruit women civilians, Secretary Baker modified his orders. Officers like Colonel Littell began cooperating with the Signal Corps at the end of October 1917 to build special housing. From Louisiana to New York, eighteen of the nation's thirty-two mobilization camps were located too far from towns for women to get back and forth easily. Civilians might not wear soldiers' insignia, but they must have housing. Promoted to Brigadier General, Littell requested $45,000 from the War Department to construct barracks just for them.[40]

In December 1917, the army modified its warnings while revealing a continuing mistrust of both sexes when placed in proximity—and deep suspicion of footloose, unmarried women. Secretary Baker ordered that females should be recruited only when males "cannot be obtained" and placed under "constant supervision." Commanders must "exercise the greatest possible care" in selecting women, hiring only those "of mature age and high moral character." Only then could females be "permitted in camps without moral injury either to themselves or to the soldiers."[41]

Newton Baker and officers like I. W. Littell were not just being ornery. They were coping with a deluge of correspondence from concerned citizens about young men's exposure to the immoral influences presumed rife on army bases. As one Illinois Christian group described their fears: "We . . . respectfully petition you to help keep our boys clean; not only from the ravages of the liquor traffic, but the scarlot [sic] woman as well. We have sincerely given you our best and we sincerely trust you will not only use them, but protect them from these forces that are more deadly that the armies of Europe, inasmuch as they destroy both body and soul."[42]

Private letters echoed the War Department's worry about camp followers and even prostitutes—"the diseased vultures that hover around the outskirts" of army bases, as one described such women. Another demanded, "Shoot the lewd woman as you would the worst German spy. . . . My neighbors are legion who think as I have written."[43] In response, President Wilson appointed a special Commission on Training Camp Activities to divert troops from such unsavory amusements, and the War Department kept a sharp eye on its female personnel.

In France, the shortage of competent switchboard operators was even graver than at home. There, poor staffing tormented army communications. As a civilian expert from the Ohio Telephone System explained the problem, women were urgently needed "to replace unsatisfactory soldier operators with the army in France." From the Brest telephone exchange on France's westernmost tip, to AEF headquarters five hundred miles inland, there was "almost complete dissatisfaction with the performance of men. There have been a few cases where men have furnished satisfactory service, but very few indeed where men have worked under pressure or in numbers calling for team work, or in exacting personal service." Men simply were not "tractable," G. R. Johnston reported. They could not put up with "criticism or abuse" by short-tempered officers as women did. Females controlled their emotions better than males. "Experience, almost without exception, has come to show the great superiority of female telephone operators."[44]

Yet officers like Johnston still hoped to minimize the use of women. Perhaps the army could induce cowardly men to bend their personalities under threat of "assignment to the firing line," Johnston wrote. Male soldiers had no choice about where they served, and some were "more feminine than some women." In an army of millions, surely enough men with sufficient motivation and suitable temperament could be found.[45]

But the problem in France went beyond running complicated switchboards under pressure. There was the special problem of toll calls in a foreign language. It would take months to convoy all the machinery needed for an autonomous AEF network. Until then, U.S. operators must interface with local operators who spoke no English in order to connect calls across the country. When Pershing and his first troops arrived in France in June 1918, the AEF leased antiquated French equipment and used former telegraph men to patch long-distance calls through rudimentary phone systems run by French women.

These conditions frustrated Pershing from the start. "Business we were accustomed to do over the telephone the French did by mail and in long hand," Pershing's biographer and companion noted. "Connection over the overworked and limited French telephone system was maddening."[46] Language barriers compounded delays occasioned by feeble foreign circuits. It was a rare American who knew French. Most possessed only a few vocabulary words and struggled to pronounce even those over crackly wires.

On duty in France, Captain Robert B. Owens of the Signal Corps made a bold recommendation in fall 1917. A Johns Hopkins classmate of General George Squier and a molecular physicist associated with Philadelphia's Franklin Institute, Owens headed intelligence and translation for the AEF. As he summarized the problem, most soldiers "had never seen a switchboard before sailing," and none "could speak French." He suggested to Brigadier General Edgar Russel that they recruit bilingual American women and "give them relative military rank corresponding to the British scheme."[47]

Bemused, General Russel retorted that he "would not be responsible for the care of a lot of American girls in France under war conditions." The unconventional Owens, a reservist and former university professor, quickly became the butt of jokes at AEF headquarters.

But Colonel Parker Hitt, regular army and Owens's immediate supervisor, took up the idea. In charge of communications wherever Pershing camped, Parker Hitt was a lean, lanky man with a nonconformist streak. A groundbreaking cryptologist who wrote the army's first manual on how to break secret codes, he appreciated bright women so greatly that he married one and taught her everything he knew. Genevieve Hitt of San Antonio never attended college but became an expert cryptologist. After Colonel Hitt left for Europe, Mrs. Hitt deciphered messages in his stead. In April 1918, as the war heated up, the army hired her at $1,000 per year. Genevieve found her notions of appropriate behavior changing. As she wrote to her husband's mother:

> All this seems so funny to me, at times I have to laugh. It is all so foreign to my training, to my family's old fashioned notions about what and where a woman's place in this world is, etc., yet none of these things seem to shock the family now. I suppose it is the war. I am afraid I will never be contented to sit down with out something to do, even when this war is over and we are all home again.[48]

In France, Parker Hitt revealed his openness to unusual solutions. As he later recalled, "We had a small PBX hooked in to the city system (if you could call it that), and there were one or two French girl operators on in the daytime and a soldier at night. The service was terrible." When Russel and Hitt visited the front, they inspected British telephone systems operated by men. Those systems were "none too good," either.

The AEF needed American women, Hitt decided, regardless of propriety. "Both General Russel and I appreciated the possible complications," he admitted, "but we knew we could get service from them and service we had to have." It might be hard to convince Pershing, but Hitt imagined "the C. in C. [Commander in Chief] was sick enough by that time of French *telephonistes* and soldier operators. I know I was."[49]

Signal Corps reports emphasized the stakes that dwarfed any inconveniences. "The importance of intercommunication in warfare can not well be exaggerated," Brigadier General George Squier wrote. "The element of time is a controlling one in strategy and tactics." Without communications for even an hour, "the whole military machine would collapse." The Signal Corps needed good telephone service, which was superior to all other technologies "for the fundamental reason that . . . it permits the language directly in signaling."[50] And for that system to function, the corps needed specialists.

In November 1917, Pershing accepted the reality that French operators and monolingual male soldiers were "hopelessly inadequate" in keeping him connected with his growing legions or foreign Allies. Pershing borrowed a few switchboard operators from Queen Mary's WAAC, but Britain could not spare them for long. Accepting the advice of Owens, Hitt, and Russel, General Pershing decided the U.S. Army needed an equivalent.[51]

Pershing had acquired a reputation for rigidity during his tenure as a West Point instructor, but the career soldier also had an unusually robust immunity to dysfunctional prejudices, making him more flexible than is acknowledged by some historians, who consider him an organization man and a tad dull.[52] Pershing's father had been postmaster of a small

Missouri town during the Civil War. Opposed to secession and slavery, he insisted on flying the Union flag even though Rebel irregulars assaulted their home.

The family's fortunes went into decline in the economically volatile 1870s. After John graduated high school in 1878, he accepted a position teaching at a black elementary school. When white bullies burst into his classroom with cries of "Nigger! Nigger!" the athletic Pershing replied that Lincoln had believed "negroes had a right to education" and the men had best skedaddle. Years later, Pershing endured similar taunts after leading the Tenth Cavalry, whose African American troops he praised as courageous. Knowing this, disgruntled West Point cadets dubbed him "Nigger Jack." As Pershing rose in fame, the insult was commuted to "Black Jack."[53]

In Europe, Pershing decided he must draw from another controversial labor pool. Sweeping structural changes during the early twentieth century had vastly expanded women's participation in the workforce, guaranteeing that Wilson's wartime administration must turn to women for both civilian and military tasks. Because of sex segregation in employment, particularly telecommunications, women possessed certain skill sets that men simply did not.[54]

Captain Owens and Colonel Hitt drafted the order that Pershing sent with his daily cablegram from AEF headquarters in Chaumont, a northeastern town near the front in the Lorraine region. The November 8, 1917, cable read: "On account of the great difficulty of obtaining properly qualified men, request organization and dispatch to France of force of Woman telephone operators all speaking French and English equally well." Pershing's message dictated precise staffing requirements: three chief operators, nine supervising operators, seventy-eight local and long-distance operators, and ten substitutes.

In civilian switchboards, "chief" was the highest position normally held by females. They typically reported to male "traffic supervisors." (Colorado traffic supervisor Merle Egan was an exception that proved the rule.) Pershing's cablegram evoked this model, similar to his plan for African American soldiers who also served under white male officers. Pershing told the War Department that he needed just "one man with operating experience to generally supervise traffic." This man, he said, should "be [a] commissioned Captain."[55]

Captain Robert Owens drafted the message to make clear that operators would be full-fledged participants in the army, as he erroneously supposed the British WAACs were. Visual evidence supported the assumption that WAACs were regular soldiers. They wore khaki and answered to orders.

Against Owens's recommendation, however, Pershing's November order was modified to emphasize a more ambiguous status akin to nurses. An official army unit since 1901, nurses had no military rank other than "nurse," known as a relative rank. Like men, they served under oath. Indeed, this was Pershing's standing practice. Of the multiple experts recruited for the emergency, "all must be in uniform, all take the soldier oath." Telephone personnel would bear the rank "operator," Signal Service at Large, AEF.[56]

Like the British War Office, General Pershing (or perhaps Edgar Russel) drew back silently from the fullest implications of the army's need for women. Pershing asked the War Department to recruit females but did not request the same military status accorded yeomanettes and marinettes. Perhaps Pershing thought such specificity unnecessary. Why would the army treat females differently from the navy and the marines?

Black Jack later praised his telephone team in terms that equated men and women: "The officers and men and *the young women* of the Signal Corps have performed their duties . . . with a devoted and patriotic spirit to which the perfection of our communications daily testify" (italics inserted).[57] General Pershing did not know that America's newest soldiers would face a long battle after the war ended. Before this could happen, however, the army had to recruit a nucleus of females ready to abandon their privacy, risk their modesty, accept military discipline, and brave submarine attack. The War Department turned to this challenge next.

4

WE'RE GOING OVER

P ERSHING'S CABLEGRAM sparked immediate action and long-term confusion. He requested a force of uniformed operators with the same allowances as nurses. Every word of his order implied military service without stating it. U.S. legal code specified that only males could enlist in the regular army, but those responsible pounced on Pershing's larger intent: get qualified bilingual operators in uniform and ship them to France ASAP. Let bureaucrats work out the niceties later.

First Lieutenant Ernest Wessen was the officer on duty when the cablegram arrived. Enterprising, quick, and efficient, he sprang into action. "New in Washington, I asked no questions, [but I] proceeded to organize and train the units in the best manner I knew how," Wessen later wrote. He rang the New York office of Bell Telephone to arrange for training and contacted the French embassy to help test for language proficiency. Then, within a "matter of moments, I phoned the Associated and United Press Bureaus, and told them that, at last, women were to be allowed to serve overseas in and as a part of the Army."[1]

Wessen was an idiosyncratic collector of rare books, who made his first literary acquisition with the $2 his father gave him for new shoes. Unschooled in government protocol, the

imaginative reservist found out within the week that he had
offended superiors up and down the chain of command. Official requests of AT&T had to be filtered through Major John
Carty. Contacts with foreign embassies must go through the
State Department. George Creel of the Office of War Information should clear press announcements. Lieutenant Wessen
found himself on the carpet at the White House trying to provide "a good reason why I should not be shot at sunrise."[2]

The best reason, it turned out, was that Wessen had started
the ball rolling in the precise direction Pershing desired. Two
days after Black Jack sent his order, Wessen began a mailing
campaign to newspapers and postmasters around the country
advertising the army's urgent need for one hundred bilingual
operators for "immediate service in France."

They would earn the same pay as male soldiers doing the
same Signal Corps work: $60 per month (three times the
wage of naval yeomen clerks). Supervisors would earn $72,
chief operators $125. Salaries were comparable to civilian
switchboard wages, so they presented no economic incentive.
In fact, some operators took a pay cut to serve the army,
without any compensatory stipend such as AT&T gave men.[3]

In addition to wages, the army would grant billeting allowances on the same basis as nurses. Recruits would wear a
similar military uniform. All must be fully fluent in French,
and it would "be a waste of time for any but fully qualified persons to apply for these positions," Lieutenant Wessen's press
release emphasized. One phrase stood out, in a separate para-
graph of its own: "Employment, of course, will be for the duration of the war." The press release opened and closed with a
plea to newspapers to give this "the greatest publicity."[4]

The word "employment" hinted at the confusion lingering
around the edges of the unprecedented initiative. Female

switchboard operators at cantonments in the United States were clearly civilians, but General Pershing wanted women in uniform. Wessen understood this to mean that those serving in France would be "combatants under international law."[5]

Army uniforms and brass insignia conferred status. Only enlisted personnel and officers were legally entitled to wear them. They were badges of authority that protected captured soldiers from being executed as spies or criminals. As AT&T explained to women recruits, "If captured, [you] are entitled to the privileges of prisoners of war, and for any offense against military regulations are subject to court-martial."[6]

Two weeks after Wessen's announcement, the adjutant general of the United States—the army's legal counsel—sent an internal memorandum to clarify the new recruits' status: "These women will be civilian employees of the Signal Corps and will have the privileges and allowances now prescribed or which may hereafter be prescribed by Army regulations and General Orders for Army Nurses." They would be "authorized by contract and not by Army regulations." The water remained muddy, however. No one prepared contracts for the operators to sign. And, as General George Squier later explained his understanding, if the operators were like nurses, then they "served by appointment and not by enlistment" but, nonetheless, as part of the regular army.[7]

General Squier's position was understandable. Operators would handle all telephonic communications and thus national secrets. Only uniformed soldiers under military discipline occupied such positions of extraordinary trust.

The judge advocate general resisted this logic. On the same day in March 1918 that the first women steamed out of New York harbor on a camouflaged troopship, the army's top lawyers ruled that operators were ineligible for war risk insurance,

unlike female yeomen serving at home. Despite this, each operator on board wore the same mandatory "identity discs" as
every doughboy, one for nailing to a temporary marker if
she was killed, the other for burial squads to match her body
with army records.

The women were ineligible for insurance because they were
not true military, army attorneys told the Signal Corps. Congress had passed no law allowing their recruitment. "There
is no authority for the enlistment of women in the army or in
any corps or department thereof," they argued. "Women employed in the Signal Corps are civilian employees, and are
not members of the military forces of the United States
Government."[8]

Two weeks later, still in March 1918, the general counsel
for the Treasury Department's Bureau of War Risk Insurance
reiterated to the chief signal officer that the operators had a
"civil status" and were "engaged under contract."[9] Yet none of
the women on their way to France had signed a contract. All
had sworn the army oath, which was their only agreement
with the U.S. government. They had relative ranks: operator,
supervisor, and chief operator. Some in the first and second
units were even offered war risk insurance.[10]

The army was skirting the truth, as evidenced by its
approach to a similar problem with female physicians. Under
political pressure to use women doctors, army lawyers privately admitted to Secretary of War Newton Baker that the
National Defense Act of 1916, which had inducted men like
John Carty, did "*not* prescribe sex as a qualification for
appointment to the Officers' Reserve Corps." The word "male"
appeared nowhere in the emergency legislation. The U.S. president could appoint anyone he deemed "physically, mentally
and morally qualified." Yet women did not meet this defini-

tion according to "custom and usage," the judge advocate opined. By virtue of gender, they could never be qualified. The head of the army's War Plans Division advised Secretary Baker to dissimulate. He should tell senators and women's club members who had taken up the doctors' cause that "under the present law women physicians cannot be commissioned in the Reserve Corps and that it would take an act of Congress to bring this about." This was not true.[11]

In letters that reeked of sophistry, army lawyers ruled that females innately possessed "physical disqualifications for the work of a medical officer . . . in the Military Service." Nowhere did they describe the nature of these defects. They did say women could perform such jobs in a civilian capacity, however. The service would allow female physicians to work on a contract basis as civilian laboratory technicians, medical aides, and nursing instructors. The most highly trained might be contract surgeons, but no one would salute them. In other words, the army simply wished to deny women the status of soldiers. They could draw government paychecks, but not as officers or enlisted personnel.[12]

Ernest Wessen was not privy to the lawyers' shenanigans and may not have examined the 1916 National Defense Act. He considered the judge advocate general's self-contradictory policy with regard to telephone operators "a fool thing which I did not choose to argue at the time." He was following Pershing's instructions and would sort out the enlistment problem later. What Pershing wanted, Pershing got. As Secretary of War Baker told the AEF commander before he sailed for France, "I shall give you only two orders, one to go and one to return."[13]

Wessen also had no choice. Women possessed certain exceptional qualifications. As he told reporters, "It would be impossible to brigade a troop without these girls." Quoted in newspapers across the country, Wessen promised, "They are going to astound the people over there by the efficiency of their work. In Paris it takes from forty to sixty seconds to complete one telephone call. Our girls are equipped to handle 300 calls per hour." Women could connect roughly five calls in the time it took a man to complete one.[14]

Lieutenant Wessen did not doubt that bureaucrats would eventually see reason. Most of his superiors had already forgiven him for going over their heads.[15] But Wessen underestimated the army's deep reluctance. Stateside, in the safe environs of Washington, D.C., army leaders remained far more inflexible than the navy. Unlike Josephus Daniels, Newton Baker opposed any modification of protocol.

Curiously, Baker was a supporter of women's suffrage. As Cleveland city attorney in 1911, he had argued, "There has never been a more stupid blunder made in the world than where the cleavage was made—putting men on one side and women on the other in the conduct of the world's affairs." In contrast with the more conservative Wilson during the same period, Baker believed "we are going to have women's suffrage of course . . . brought about by economic conditions." Presumably, he meant conditions like the nationwide employment of females as operators of the world's most advanced telecommunications system.[16]

Baker expressed himself on women's suffrage even more strongly once he became secretary of war: "In 1789 it might well have been possible to define a democracy as a society in which the family was represented by a single representative—a man; but in 1917, society cannot speak of itself as a democracy unless . . . all the men and women who live under the administration of that government and those institutions are recognized and represented in the Government." He especially commended women's wartime contributions in the civilian economy. If they were "to stop tonight doing the things that they are doing . . . we should have to withdraw from the war."[17]

Nonetheless, Newton Baker's acceptance of a new political role for women did not prompt him to set out a welcome mat in the army. When members of Congress introduced a bill in December 1917 to enlist women, as the navy and marines had already done, Secretary Baker quietly nixed the motion the day after Christmas. "The enlistment of women in the military forces of the United States has never been seriously contemplated and such enlistment is considered unwise and highly undesirable," he wrote the chairman of the House Committee on Military Affairs.[18] Congress did not pass legislation to enlist women because Baker opposed it. Telephone operators and female physicians were not inducted into the Officers' Reserve Corps for the same reason.

Perhaps Baker felt the need to defend his own virility. A small man, not much above five feet, the bespectacled secretary often sat with one leg bent under him to give an impression of greater height when receiving visitors at his giant desk in the War Department. Baker battled partisan critics who wanted to know why the United States had done so little fighting since entering the war nine months earlier. Accused of being a "chattering ex-Pacifist," Baker had to explain himself more than once to the skeptical Senate. In the press, critics called for a "he-man" to fight a "he war." Or, as some said, "We need a butcher, not a Baker."[19]

Whatever Baker's reasons, the Hello Girls lacked a civilian champion on the order of Josephus Daniels. They had the operational support of General Pershing, who needed them desperately, but the War Department was scared of female soldiers.

The press was unaware of the controversy. So were recruits, who thought they would be enlisted. It was a thrilling prospect at the start of the twentieth century, when commentators spoke often of the "New Woman." While officials quietly skirmished, the public latched on to Wessen's message: the government was recruiting females.

"Army Wants Women," the *New Orleans Times-Picayune* told readers in November. The government would uniform and compensate them on the same basis as nurses. "The Army Calls Telephone Girls," the *Kansas City Star* informed townsfolk in December. "Women Phone Operators for Army," the *Philadelphia Inquirer* echoed. "They have the same privileges and allowances as Army nurses." The *Duluth News-Tribune* reported "Pershing Forces to Need Hello Girls and 20,000 Nurses." The switchboard force would number 150 "physically fit" bilingual "young ladies." They would be "enlisted for the duration of the war," with the same quarters, rations, pay, and uniforms as nurses.[20]

Bell Telephone News ran a fuller story in February titled "Young Women of America, Attention!" Alongside a "Roll of Honor" featuring telephone men overseas, the magazine told readers, "Here's your opportunity to serve your country in France with General Pershing's Expeditionary Force." The magazine gushed that women would "do as much to help win the war as the men in khaki who go 'over the top.'" Operators whose French was "very, very good" could be "switchboard soldiers." The Signal Corps sought "level-headed young women . . . able to exercise good judgment in emergencies and willing to work hard and even endure hardships if necessary." Operators would wear uniforms "at all times." Civilian clothing would be of "no use to these young women soldiers 'over there.'"[21]

Southwestern Telephone News, the Bell publication aimed at Arkansas, Kansas, Missouri, Oklahoma, and Texas, told its readers that "girl telephone operators" in the Signal Corps would be regular army. When one employee later made the force, the magazine praised her as "a full-fledged soldier under the articles of war." Sworn in at Love Field in Dallas, Pearl Baker took an oath that "subjects her completely to army discipline." Southwestern Bell posted a "word of warning," though: no woman should apply who was not completely fluent in French.[22]

In the first week of December 1917, letters began pouring into the War Department, which had not yet even printed application forms. More than seventy-six hundred women inquired about the first one hundred positions. Echoing the strains of George M. Cohan's wildly popular song "Over There," the letters expressed applicants' eagerness to "do their bit," help "send the word," and win the war "over there." Lieutenant Wessen reviewed the inquiries with an AT&T team in New York headed by one R. F. Estabrook.[23]

Wessen's committee devised a letter to potential recruits. It explained, "This will be the only unit composed of women which will actually wear Army insignia. This Unit will be very similar to the British Women's Auxilliary [*sic*] Corps . . . which has gained considerable fame in France."[24] The committee met Christmas Eve to make arrangements for interviewing applicants to assess mental acuity, physical fitness, and ability to conduct simultaneous translations. Letters went out in early January. The committee narrowed consideration to 1,750 strong prospects, some from French-speaking regions of Canada.[25]

Officials improvised as they went along. Need ran ahead of protocol, and protocol struggled to catch up. In the belief that most intelligent women could become competent at telephone operating, they focused their search on the less common qualification, bilingualism. Indeed, since telephone companies had previously avoided immigrants who spoke with an accent, recruiters now had to reach outside the ranks. Unlike the Bell Battalions, women's units could not be lifted wholly from within.

Major John Carty suggested hiring one hundred women who spoke French, then training them in telephone operating. The Engineering Department of AT&T would run as many training programs as needed across the country to bring smart bilingual women with no experience up to speed.[26]

In an analysis of the spotty personnel records still in existence a century later, historian Jill Frahm found that only 42 percent of those inducted had prior telephone experience, whereas 78 percent spoke French. Technical skills could be learned on the job; language acquisition took years.[27]

Like male volunteers, females possessed a range of motivations, from a yen for adventure to a desire to help family in the war to a hatred of German aggression. Many had ties to France and Belgium as strong as the ones they had to the United States. Multiple motivations are normal; humans are complex creatures with shifting emotions and ideas. Yet patriotism was the primary reason for most, according to their testimony. As one young woman described her emotions, "[Pershing's call had] aroused in me a great, unquenchable, patriotic desire to do my bit."[28]

The United States had not fought a major foreign war in a hundred years. The sense of alarm was electrifying. *Bell Telephone News* exhorted workers that love for the flag must be transformed into action. "This war will not be won unless every person does his or her part, gives his or her service, makes his or her sacrifice, places self behind the need of the nation. . . . It means loyalty in little daily things as well as loyalty to the great principles of our government."[29]

Not everyone responded to such pleas, but they did sway millions, including the 223 American women who eventually labored at army switchboards in Europe, a few within sight and sound of the furious battlefield. They were an intrepid group with mixed backgrounds, united in common purpose. Inspired by Joan of Arc, history's most famous female patriot-warrior, they wanted to save France.

Their torchbearer was twenty-five-year-old Grace Banker—a bright-eyed, chestnut-haired native of New Jersey, who possessed a strong sense of purpose, an extraordinary stock of energy, a sensible personality, and an open smile. Curiously, the woman who led America's first female soldiers onto foreign soil exhibited the greatest circumspection among those who responded to the army's call for volunteers to serve for the duration. Perhaps that is why she was chosen.

"I should like to join for one year with the privilege of re-enlisting," Grace Banker wrote prudently. "If that is not possible, I will enlist for the period of the war." She was the only applicant to make this request, which was ignored.[30]

It wasn't that Banker lacked gumption. She had played male roles before—in fact, twice before in the amateur theater of Barnard College, the women's institution under the sponsorship of Columbia University, which then admitted only male undergraduates. As a freshman, Grace was elected treasurer of the Church Club and won the part of First Prophet in the club's Christmas play. Louise Walker, described as "lovely" by the *Barnard Bulletin,* beat out Grace for the role of Mary, mother of God, but Grace could not have mistaken the assignment for a slight against her femininity. All other actresses had to play men, too, since Mary was the only female character in the fifteen-member cast of kings, shepherds, angels, and prophets.[31]

The next year, Grace starred as the Right Honorable Henry Carlton, M.P., in *His Excellency the Governor,* the sophomore class play. Arriving in a tropical backwater of the British Empire, the stuffy old legislator found himself ensnared by a pretty young gold digger to comic effect. The school newspaper opined that Grace Banker's performance was "well done," although she "was not quite as dignified" in her portrayal of the Right Honorable Henry Carlton as might be desired.[32]

Those she later commanded in France would not have been surprised. Grace Banker played well the part of a stern, self-possessed government officer. Yet she had a fun-loving streak underneath. Both characteristics proved vital in leading the women of the Signal Corps. Combined with a prestigious university degree, they made her officer material.

Following graduation in 1915, Grace Banker landed a job at AT&T headquarters in New York, where she advanced quickly. She read the notice regarding "recruits for telephone service in France" in the Sunday *New York Globe* on December 5, 1917. Four days later, she wrote the chief signal officer: "I understand both French and English and have a thorough knowledge of telephony." Grace waited a month, then sent another letter, fearing she had been bypassed. "If my application has not been accepted for the present would I be eligible for a second unit if such should be formed?" Her words suggest keenness but also poise: "I would appreciate any further information in regard to this matter as my plans for the future must be governed accordingly."[33]

Soon thereafter, the Signal Corps sent Banker a form asking for educational background, employment experience, medical history, and a formal photograph—or a "snap shot if good likeness." Hundreds of potential recruits received the same request. Banker may have been one of those motivated by adventure, though the diary she kept despite army regulations also makes plain her patriotism. She had a brother in the 77th Field Artillery.

Grace returned a picture of herself in a satin suit-dress with a deep white collar, hair pinned up in a bun. She stands on the lawn outside her home in Passaic, New Jersey. Her hands are clasped behind her back as if she is unsure what to do with them, but she looks calmly into the camera.

Banker was one of thousands vying for the unusual opportunity, and she may have stretched the truth about her language fluency. Another applicant in the pool considered Banker's French rusty. Louise LeBreton was a tall, headstrong student at the University of California at Berkeley and a part-time secretary for the French consulate. She had family overseas, some in the trenches. Born in France, Louise was bilingual.[34]

Yet the eighteen-year-old LeBreton stretched the truth in a different direction. She and her younger sister Raymonde read about the opportunity with the Signal Corps in the student newspaper, the *Daily Californian*, and immediately wrote the War Department. The photographs they mailed show beautiful young women with tresses done up Gibson-girl style.[35]

When the LeBretons received telegrams in early January 1918 summoning them to an interview, Louise LeBreton told the AT&T representative that she was twenty-one. He reported that Louise was a young woman of "character beyond reproach" and "intelligence of a high order," who "looks and acts older than her age." Raymonde told her interviewer, yet another gentleman, that she was eighteen, though she was sixteen. He, too, was fooled. His evaluation described the Berkeley Business College student as "considerably more mature than age would indicate" and capable of "standing up to severe strain as she is wirey [sic], well knit and alert." Both passed inspection as women of "strong character" and quick wits.[36]

The girls were not entirely truthful to their mother either, who did not know they had applied. "We were sure," Louise admitted, that she "would give us a categorical 'NO.'"[37] The LeBretons' father had died of yellow fever in Panama, a victim of the failed French attempt to build a canal. His widow immigrated to California, where she counted on her eldest girls' wages and housekeeping help with running a boardinghouse for students learning French. Jeanne LeBreton was a struggling single mother of four daughters, two younger than Louise and Raymonde.

Madame LeBreton was indeed horrified to learn of the girls' plan after the fact. She worried for the land of her birth but for her daughters more. Parental authority over unmarried females remained the custom, and the army required parents' permission even if a woman did not live at home, as the majority of applicants did not. Madame LeBreton ultimately gave her consent with the proviso that Louise and Raymonde be kept together. AT&T agreed, and the army split them up their second day in Paris.[38]

The LeBretons were not the only applicants to fudge their age. Hope Kervin of San Francisco was particularly impatient to enlist. In December 1917, she wrote the War Department almost weekly, promising to make herself an "awful nuisance" until they took her.

Kervin's letters reveal a love of France, a spirit of adventure, and a desire to get her own "little whack at a German." She knew Belgium and France intimately, having attended school in both countries. She possessed a bachelor's degree, "could handle an automobile," and expected her aviator's license before Christmas. On the eve of the holiday, she inquired if she could join the British Women's Army Auxiliary Corps, then hop over to the Signal Corps once the slower-moving U.S. unit got off the ground. Kervin followed up her saucy letters with a telegram: "If constant reminding can help I am taking no chances of being overlooked."

Like the LeBretons, Hope Kervin feared rejection, since press announcements did not specify age requirements. "I lied to get in," she admitted candidly to officials when applying for a copy of her discharge papers in 1928. "I feared they would not take us unless we were twenty-three or twenty-four. I am telling the truth now."[39]

She had reason for concern. In internal documents, the army and AT&T agreed to select applicants "without any reference to age" but preferred recruits between twenty-three and thirty-three. "The more desirable applicants will, ordinarily, be from 23 to 28 years of age and the higher limit of 33 is set chiefly to permit the inclusion of applicants with telephone experience who have advanced beyond the rank of operator."[40]

In Kervin's original application, the twenty-one-year-old claimed she was twenty-three. Her photograph shows bobbed hair and a pleated tunic reminiscent of dancer Isadora Duncan, another free-spirited Californian of the era. Despite her animated tone, Kervin looks at the camera with ennui. Perhaps she thought a world-weary expression would make her look older. Of all applicants, Kervin most explicitly projected the insouciant, independent image of the New Woman, sometimes taken as a feminist ideal.[41]

Many women simply wished to stop the killing of family and friends in Europe. The *Telephone Review*, another Bell publication, quoted a woman in March 1918 who grew tired of her interviewer's questioning attitude. She finally sprang to her feet and thumped the table with both fists. "You ask me why I want to go to France, after what I have told you the Germans did to my home and my family! I would go a thousand times, even though I knew the day I landed in France would be my last!"[42]

Recommendations for one recruit emphasized her Belgian birth. A graduate of Columbia with a master's degree in mathematics, Belgian immigrant Madeleine Batta was called brilliant, alert, and "fearless." Louise LeBreton also harbored sympathies for France's northern neighbor. "We wanted to fight for freedom and democracy," LeBreton later recalled, "and the fight the French and British put up [on behalf of Belgium] inspired us."[43]

Others possessed a keen sense of duty. Isabelle Villiers—an operator whom R. F. Estabrook personally recommended to Ernest Wessen—was already a yeoman (F) first class in the Boston Naval Yard. "In requesting this duty I feel I shall be able to serve my country to better advantage than I am doing at present," Villiers explained, attaching a photograph of herself in a white sailor's suit and hat. "Although the position I now hold is one of much responsibility, I feel that there are more who could fill it than fill the one for which I am an applicant." The navy released Villiers at Ernest Wessen's request.[44]

Others simply took for granted that females had as profound a civic duty as males. Their conviction that women must act as full citizens reflected broader social changes. Almost 90 percent of the Hello Girls had worked outside the home before joining the army. North Dakota operator Marie Gagnon was one of those unwilling to ask for special dispensation as the weaker sex. She had brothers in the Marine Corps and the army, and "wishes to do her share," the interviewer reported. Gagnon scorned shirkers. She had broken up "with a young man because he had no reason for not entering the service and was waiting for the draft." Only five feet tall, Gagnon told interviewers she was "willing to accept any conditions that the service may bring about."[45]

Merle Egan was not among the first applicants. She was a rule follower who took the language requirement seriously. Egan consoled herself by volunteering at home. She landed on the County Roll of Honor for her Red Cross work and was so successful at selling war bonds that fellow employees elected her president of the Telephone War Savings Society, the third largest of Helena's twenty-three businesses helping the government. Merle regretted that the local Bell affiliate did not become a "100 percenter," as many companies did, though 74 percent of local telephone workers purchased bonds under her determined leadership. Egan finally got her chance to go overseas when the Signal Corps relented on its bilingual requirement as American-built circuits began to predominate in summer 1918 and operators needed to connect less frequently with French switchboards.[46]

Once Wessen and Estabrook determined how to process applicants, events moved quickly for the first three contingents.

Grace Banker checked back anxiously with the Signal Corps on January 7, 1918, concerned she might miss out. Eight days later, she was ready for fingerprinting and inoculation. Banker's personnel file does not contain the supplemental recommendations that other recruits sent. Instead, the chief instructor for New York City—a woman named Louise Barbour, who soon decided to enlist as well—wrote two brief sentences on Banker's application form: "Miss Banker has been a member of the Instructor Department for the American Telephone and Telegraph Company for the past two years. I should be sorry to lose her services."[47]

To her surprise, Grace Banker was selected to lead the pioneer group. She would not outrank chief operators in subsequent units, but her contingent would be the test case. The morning Banker sailed from New York two months later, the faces of her team "glued to port holes and doors" as the Statue of Liberty drifted from sight, she confided trepidation to her diary: "For the first time suddenly I realized what a responsibility I have on my young shoulders." Nonetheless, on January 15, 1918, she raised her right hand and swore to "support and defend the Constitution of the United States against all Enemies, Foreign and Domestic."[48]

The following day, on January 16, R. F. Estabrook of AT&T informed Wessen, now a captain, that he had appointed Mrs. Inez Crittenden of San Francisco to head the second unit. Crittenden was a divorcée, then a social handicap, but she was more mature than many other applicants and possessed supervisorial experience. Crittenden took the initiative of submitting a recommendation before being asked. Her employer vouched for her in December 1917: "She is well educated, speaks French, is industrious, thoroughly reliable in every way, and capable of filling well any position she may apply for."[49]

The choice of Inez Crittenden as the second chief operator indicates how frantic the Signal Corps and its AT&T partner were for experienced, bilingual women to act as officers. Crittenden had no prior telephone training, but she had a commanding presence. Like a few other volunteers, Crittenden purchased war risk insurance before sailing, unaware that she was ineligible. Tragically, that determination was made later.[50]

The Signal Corps meanwhile turned to selecting the women whom Banker, Crittenden, and subsequent chief operators would command. Evaluators sorted through pictures and recommendations sent by applicants from California to Maine, perhaps unsure what a "level-headed switchboard soldier" looked like. Many women attached formal portraits taken in photographers' studios. Some wore sensible suits and hats. Others chose lacy, high-necked Victorian blouses. One posed in her garden with a pet rabbit, another sat astride a pony, a third perched sportily on a tree branch.[51]

Recommendations typically praised applicants' loyalty, a matter of such great concern that the War Department required at least four testaments to candidates' patriotism. Operators could overhear every military telephone conversation. They would be translating orders between French and American officers. Intelligence would pass constantly over the wires in their hands. As the Associated Press explained to readers nationwide, "The work to be performed was largely of a confidential nature and would give the operators carefully guarded information as to troop movements." Recruits would be examined by psychologists "using methods employed by the army in judging the qualifications of officers."[52]

The letters regarding Louise Barbour were characteristic. Recommenders applauded the Cincinnati native as "good old

American stock," "thoroughly patriotic," and "impelled solely by her desire to do all she may in the battle to save civilization." Women with foreign-sounding names were described in compensatory terms as "100 percent American." Each applicant swore a loyalty oath upon induction. Many took a second oath upon transfer and a third upon promotion.[53]

Merle Egan pledged herself twice before training. The adjutant general for Montana recorded her first oath, then a telephone company representative. The former told her, "This is the first time that I have sworn a woman into the Army, but hold up your right hand."[54]

Applicants of German or Jewish backgrounds got a closer look. Helen Orb of Chicago—a talented sculptor studying at the École des Beaux-Arts in Paris when war broke out—was one of the first to apply. Her German father was a naturalized citizen who had lived in the United States for fifty years. AT&T undertook special investigations to determine if the family was "entirely loyal to the United States."[55] Eventually satisfied, they sent Orb to Europe under Crittenden's command.

Minnie Goldman prompted a more thorough investigation. A graduate of the University of Chicago, Goldman worked as a telephone operator before attending Northwestern Law School. A member of the Chicago Bar Association when she applied for the Signal Corps, Goldman was initially rejected for insufficient fluency in French. The army also suspected her family of being pro-German. "While I have no occasion to doubt her loyalty," a Bell interviewer wrote, "I also have no positive evidence of it; and she is partially German Jewish."[56]

Goldman persevered. She polished her French for another two months and wrote Captain Wessen to clarify that her Jewish parents were "of Roumanian origin." They had fled Europe to avoid massacre. AT&T continued its investigation, questioning employers and acquaintances. One even pumped Goldman's mother, who "intimated that her neighbors are not as loyal as they might be, and said that she only wished the Federal authorities knew the feeling of her neighbors as she does." The investigator left convinced of Goldman's loyalty, since family values were "usually reflected through the wife, especially among this race of people."[57]

Anxious to find a sufficient number of bilingual operators, the Signal Corps took at least one applicant of questionable maturity but undoubted fluency. Lucienne Bigou was forthright about being eighteen, and a recommender praised her as honest, patriotic, and "very bright." But was it fair, Constance Ball asked the phone company, "to subject a girl under 21 years of age to these conditions?" Lucienne Bigou was competent, but Ball felt she was "too young to be put face to face with temptations and dangers that have bewildered many older women." As events evolved, Constance Ball proved prescient.[58]

Interviewers scrutinized recruits for bravery as well. Cordelia Dupuis, a North Dakotan who sailed with Grace Banker, was twenty-one but showed a "keen desire" to serve, appeared self-reliant "beyond her years," and seemed "perfectly willing to meet any conditions that the service may bring about." Dupuis demonstrated an "absolute absence of nerves," her interviewer commented. Merle Egan's recommenders similarly described the Montanan as a "loyal true American" who "never showed signs of nervousness."[59]

Overall, it was a remarkable applicant pool. Telephone operating was one of the better white-collar fields open to women, making it fiercely competitive. Applicants to the Signal Corps possessed an extra degree of initiative: they were willing to do most anything to serve their country. Only Captain Ernest Wessen stood in their way—or could open the door.

5

PACK YOUR KIT

T HE AEF urgently needed women to run telecommunications, but only the best would suffice. In two months or less, the pioneers had to be selected, trained, outfitted, and dispatched. Grace Banker's group was set at thirty-three, Crittenden's at forty-two. Louise and Raymonde LeBreton sailed with the first unit on March 2, 1918. Impatient Hope Kervin had to wait for the third, which left in April. Merle Egan sailed with the fifth unit in August, on the same troopship as Louise Barbour, when the Signal Corps realized it needed hundreds more women once their performance proved outstanding.

Before operators made the final cut, they were summoned to six training centers: New York City, Scranton, Trenton, Chicago, San Francisco, and Lowell. The telegram Merle Egan received was typical: "Under authority of Secretary of War . . . you will proceed to New York City, reporting your arrival to R. F. Estabrook. . . . Travel directed is necessary in the Military Service."[1] Next came language tests, medical exams, and telephone training.

Written testaments about fluency in French were insufficient to confirm their place. The first applicants also had to pass muster with tough examiners. The French exam challenged even native speakers, especially those with limited telephone experience. Applicants waited in line up to an hour, then took a seat in a room with only a desk and a phone. Without forewarning, the phone rang. Surprised applicants typically answered after a moment. A Signal Corps officer chronicled what came next:

> "Hello," comes a strange voice into the receiver. "This is the Adjutant, ——th Division, located at ——, in France. The Commanding Officer, General Jones, wants to talk to Colonel La Roux of the ——th Brigade, ——th French Army Corps, located at ——. General Jones understands no French, and Colonel La Roux understands no English. It will be necessary for you to translate General Jones' message into French and Colonel La Roux's reply into English. Ready? Hello, this is General Jones now speaking . . ."[2]

The playacting would then switch rapidly from French to English in a long, complicated conversation while evaluators listened in. Candidates were judged for hesitation, confusion, imprecise translation, and inapt pronunciation of either English or French. In a group of ninety-three women at Northwestern Bell, only five passed.[3]

Recruits also had to provide evidence of a private medical evaluation, followed by an army exam. Most local doctors sent routine reports on blood pressure, height, and weight. Some did basic diagnostic tests on urine and blood. Grace Banker's doctor answered yes and no to a few simple questions about her digestion, nerves, and general health.

A few women endured more grueling preliminary exams, which testified to their resolve. Janet Jones of Ohio was pursuing a master's in French at Columbia when she applied. Her father had served with distinction in the U.S. Civil War.

Jones's doctor not only tested her urine and blood but also evaluated her reflexes, ears, heart, abdomen, teeth, and scalp. He stopped at giving the unmarried woman a vaginal examination but reported that a "rectal examination shows no evidence of hemorrhoids."[4]

Undaunted trainees underwent yet another complete physical examination by army doctors, who vaccinated them against disease—a procedure not yet routine among the general public. One Signal Corps bulletin reported, "Out of 60 girls who were inoculated . . . not one fainted. An officer who has seen many soldiers meet the same experience said this was most unusual."[5]

But medical procedures in any military setting could prove hazardous. Anita Brown was a Los Angeles native and graduate of Wellesley College, whose interviewer characterized her as a patriotic French teacher with a "brilliant mind and character above reproach." When Brown arrived in San Francisco in April 1918, the army surgeon turned her over to his assistant for inoculation. Sergeant Joseph Verdi gave Brown a shot, then wiped the length of her arm twice with a solution he assumed was rubbing alcohol.

When Brown complained of burning, the sergeant realized he had applied an acid used in minute quantities to remove warts. He apologized profusely, telling Brown that he had been thinking about his ferryboat commute to Berkeley and "ought to be shot for causing such suffering." Sergeant Verdi hurried from the room and came back with a weak alkali to neutralize the acid. He neglected to report the incident to his supervisor.

Brown returned to her quarters but experienced pain "so blinding that I was unable to reach my hotel five blocks distant." She stopped at a nickelodeon to compose herself, then staggered on thirty minutes later.

Over the next few months, Anita Brown persevered with grueling training that required constant use of her arms. Though the wound refused to heal, she traveled with other operators to the New Jersey port of embarkation, hoping to make it to France. But the open sore became progressively worse, and further army medical attention was so poor that she was unable to sail. Brown was discharged with severe, permanent nerve damage. The judge advocate general recommended that Sergeant Verdi be court-martialed for "gross and inexcusable negligence."[6]

Marie Josephine Hess was another recruit whose Signal Corps career was impeded by illness. A paid organizer for the Woman Suffrage Party of New York, Hess had lived in a French-speaking region of Switzerland until age ten, then moved to the United States, where she attended Wellesley and Cornell. Unlike the British Pankhursts, Hess had no need to suspend her suffrage organizing during the war. New York passed a state amendment on November 6, 1917, two days before Pershing's telegram. But Hess had a bad reaction to the army's typhoid inoculation. This delayed embarkation for France, though she had been initially certified as "physically qualified for foreign service." The army took no chances. After devoting nine months to the Signal Corps, Hess was still in New Jersey awaiting orders on the day of the armistice.[7]

Recruits who passed language tests and medical exams had to clear one more hurdle before transfer to New York to undergo final processing. Since many had limited telephone experience, AT&T put them through a course of intensive instruction in PBX systems, magneto boards, and common battery exchanges to learn local and long-distance techniques. Recruits received daily orders to report to different exchanges in unfamiliar cities, and they were expected to obey promptly and learn quickly. Approximately 15 percent flunked this final phase.[8]

Nonetheless, those suited to the work at last had a chance to begin enjoying their adventure. These recruits met interesting new acquaintances, saw new towns, and mastered important skills. Two whose letters have been preserved were Adele and Eleanor Hoppock of Seattle. The sisters majored in French at the University of Washington, which sent at least eleven students to the Signal Corps and graduated them early to speed them on their way. The Hoppocks traveled to San Francisco for training. Twenty-five percent of all recruits came from the West Coast. California had the greatest number of operators from any one state. Leading the entire world in telephone use, San Francisco was the logical locale for the only training center west of the Mississippi.[9]

Eager to dispel the notion that women soldiers were unfeminine, heavyset amazons, Adele Hoppock's college newspaper described the senior as "small and dainty . . . with big blue eyes." She traveled ahead of her younger sister in February 1918. Other students from Seattle met the train. They swapped stories excitedly. Adele Hoppock's "buddies" (World War I slang for "GI") nicknamed her "Hoppie." The next morning, Adele was examined by the same army surgeon as Anita Brown and began a three-day series of vaccinations. "So you can see I am in for it all right," she wrote her mother.[10]

Adele dove into training that afternoon. She practiced first on a dummy board, then on local lines. PBX training followed. Within five days she was sent "out of town" to the East Bay town of Richmond, a jitney ride from the YWCA on Nob Hill to the ferry terminal on Market Street, then across choppy San Francisco Bay on a boat. Warned about detectives searching for German spies, Adele watched what she said publicly. Awaiting more students from the University of Washington, she wrote her mother to "tell the five girls to study French like the dickens. I will be sent soon over to M. Chinard at the U [University of California] and will be given a stiff exam in French. Br-r-r-r-!!!!"

The women reveled in one another's companionship. They laughed and chatted and had a "gay time" sharing rooms at the YWCA until one by one the recruits voluntarily decamped to find quarters with French families. Marjorie McKillop, whose brother took a German machine gun bullet in the thigh a few months later, was first to announce "she felt the need of improving her French" and was ready to endure some inconveniences. Feeling "very lonesome," Adele took quarters with the cousin of a friend of a friend, who would therefore "not be exactly a stranger." The girls found a Parisian tutor to give them additional language lessons nightly.[11]

Within a few days, a supervisor told Adele "she had orders from headquarters to speed [her] up." The "whole affair seems peculiar to me," Adele wrote her mother and father, whom she called "Muzzie and Puzzie." Then she learned why. The unit might be sent to France "at a moment's notice." Adele passed the language exam without difficulty but, to her disappointment, soon heard she must wait for the next unit. "It was quite a blow," Adele confessed, watching other recruits pull out for New York without her.[12]

Another impatient trainee was Berthe Hunt, one of only eleven married women in the Signal Corps. Hunt was thirty-three and the eldest daughter of yet another French woman in the Bay Area raising four children alone. Like Louise

LeBreton, Berthe helped support her family before marriage. Embarrassed by hand-me-down clothes as a high school student, Berthe worked hard to get into Berkeley. A younger sister recalled Berthe as undemonstrative yet loving, thrifty yet generous, strict about rules yet discreet about mischief. Outwardly, the bespectacled married woman appeared phlegmatic. Inside, she was high spirited and emotional. She also found telephone work hard. The training in San Francisco was "no snap," and the inoculations made her achy and feverish.[13]

But her husband's safety was Berthe's paramount concern. Reuben Hunt, whom Berthe affectionately called "Rube" or "my boy," was a navy physician aboard the U.S.S. *Moccasin,* a confiscated German ship that was part of the food convoy to France. Rube and Berthe had known each other since high school. When Rube got orders to sail, the couple made a pact to keep diaries while separated, even though it was forbidden. They ignored regulations in yet another fashion, too. Wives of officers were ineligible to serve, but Berthe Hunt somehow wiggled in.[14]

The couple said good-bye two weeks before Berthe received her acceptance from the Signal Corps. Rube's departure for New York was a "red letter, 'rainy' day." Perhaps "rainy" meant tearful. But Berthe's spirits revived when she took the army oath to defend her country: "a proud day!" Like Adele Hoppock, Berthe learned she must await a later contingent. The postponement meant she and Rube would not see each other in New York before he risked the dangerous Atlantic. "What a disappointment!" Berthe confided to her diary. A few weeks later, on March 22, she wrote, "Last letter from my boy dated this day." Hunt wondered if she would ever see him again.[15]

For Adele Hoppock, the delay meant a reunion rather than a separation. She remained in San Francisco long enough to welcome her younger sister Eleanor, who had been practicing on a switchboard in Seattle. Eleanor found cosmopolitan "Frisco" lovely. The sisters shared Adele's room with a French family and petitioned their supervisor to serve together. He punctured their optimism. Like the LeBreton sisters and most sibling pairs, of which there were at least fifteen, they were soon separated once again.[16]

Adele Hoppock and Berthe Hunt finally received military orders for New York. A few days before leaving, they and twelve other operators rode in a Liberty Loan Parade, waving to cheering crowds from three open-air cars emblazoned with banners stating, "We are on our way to France to serve our country." A woman dressed as Joan of Arc rode a snow-white horse alongside, "linking the present with the past," *Pacific Telephone Magazine* reported, "and symbolizing the spirit and loyalty of women in the struggle for the triumph of the right."[17]

On April 9, the women boarded a train with "tearful farewells" and "enough candy & fruit for a village," according to Hunt. AT&T placed the married woman in charge of the group that, to her dismay, included "our two flirts." Navy officers traveling on the same train shadowed the women, receiving little discouragement from at least some of the alluringly named "Hello Girls." Hunt felt compelled to shoo them away. "I made myself very much in evidence & hung around so there was no trouble." Nonetheless, Hunt enjoyed sights like the Great Salt Lake, a "veritable ocean with naval & whitecaps," and the snowcapped mountains of Utah as the train chugged eastward to Chicago to pick up more operators, then on to New York.[18]

Despite her separation from Adele, Eleanor remained cheerful. She felt proud to join the Signal Corps and wrote home, "I am no longer Miss Hoppock, but Operator Eleanor

Hoppock." The Seattle co-ed's commitment to soldiering stopped at her appearance, though. "I took the fatal plunge and had my hair waved," she wrote after Adele left. "It took almost three hours and was done by a woman who had done Mary Pickford's twice. . . . It certainly looks nice."[19]

The younger Hoppock was not flighty, however. The Signal Corps promoted Eleanor to supervisor in the fourth unit, and she took the oath of allegiance yet again. Her group departed for New York soon thereafter, with renewed cautions not to "speak freely of our mission to strangers."[20]

From wherever they traveled, the operators cohered as military units in Manhattan. There they met their chief operators, got fingerprinted and photographed for passports, heard lectures on "hygiene," practiced drilling, and listened to talks by Signal Corps officers, including Captain Ernest Wessen. The army issued them a list of items to purchase, from tan parade gloves, woolen stockings, sateen bloomers, and union suits, to bicarbonate of soda, Lysol, iodine, and sewing kits.

Uniform coats, skirts, blouses, hats, high-top shoes, and bronze insignia were the most expensive items. The entire kit cost between $250 and $350. AT&T loaned money to those who could not afford the cost, with regular installments deducted from their salaries. For many earning $60 monthly, half a year's salary went into their kit. (Some served barely long enough to earn $350, meaning they received virtually no compensation once the debt was repaid.) Operators were told their status was similar to that of officers, so uniforms must be individually fitted. By contrast, yeomanettes received standard issue suits at no cost.[21]

Yet the women loved the uniforms, which meant full membership in the American Expeditionary Forces. As the *Telephone Review* of New York described the "Girls in Blue" to curious readers, "Every soldier must have a uniform, and the telephone operator units, being part of the Army organization, are no exception to the rule." Women in the garb of the American armed services "will have in their hands the wires that serve as the nerves of our army in France."[22]

Berthe Hunt expected to feel "terribly self-conscious in it, but wasn't a bit." Only the high collar was "deucedly uncomfortable." Adele Hoppock worried she might feel conspicuous, but the uniforms were "pretty and neat," she reassured Eleanor, and "anything but flashy." Signal Corps regalia soon became "quite an accustomed thing," she told her sister. "We feel quite proud, but I assure you, one gets used to it and doesn't think anything about it."[23] The women mailed their civilian clothes home, as ordered.

In New York, some operators initially roomed two to a suite at the Prince George on Twenty-Eighth Street, the first hotel in the city to provide private baths in every room. Berthe Hunt and her companion were thrilled. After a week, most shifted to the port of embarkation at Hoboken or into temporary barracks at the Young Women's Christian Association (YWCA) in Manhattan. Grace Banker and the LeBreton sisters' were billeted in a Hoboken bar, thirty-three army cots flanked by a trunk and a small suitcase. Berthe Hunt was fortunate to draw lots for a single room at the Y. Sixteen women in the third unit shared a garret of the YWCA. Yet Adele Hoppock did not mind. "We have awfully good times," she wrote Eleanor.[24]

They also learned military protocol. The female Signal Corps was under the supervision of Captain Wessen and reserve officers at AT&T headquarters while in the United States but needed female leadership as well. The unprecedented situation required improvisation. Like Major John Carty, none of the women had had officer training.

The positions of operator, supervising operator, and chief operator—which paralleled civilian roles in the urban telephone exchanges—became their ranks, which they wore as cloth insignia on their sleeves. The outline of a telephone transmitter embroidered on a white brassard, or armband, indicated operator. A laurel wreath below the transmitter meant supervisor. A lightning bolt above the transmitter indicated chief operator.[25]

"Operators must obey, without question, the orders issued by chief operators, supervisors or operators in charge, who are in authority over them," AT&T instructions explained. "Provision is made for recourse in case orders are unjust but . . . all orders must be obeyed first and immediately as required in accordance with military discipline." Operators saluted supervisors and so on, up the chain of command.[26]

Chief operators themselves were tested on protocol, sometimes without warning. One morning, while the first unit was confined to the port of embarkation awaiting sailing, an unfamiliar officer approached Grace Banker. He asked her when she was set to sail. She replied, "I don't know," though she did. He laughed and responded, "Well, of course you really do. So do I. So it's quite all right to tell me. When do you sail?"

She replied, "Well, if you know, then there is no need for me to tell you."

Later that day, Banker encountered the same man at the censor's office, to which she was delivering her unit's mail for clearance. He grinned and explained that he was with G2, the intelligence section. "Just as well you wouldn't tell me," he said. "I was trying you out."[27]

The Signal Corps further improvised a secondary hierarchy within the units, since there were too few designated super-

visors. Each new unit was divided into groups of ten upon arrival in New York, with one operator below the level of supervisor training others. Twenty-year-old Adele Hoppock at first
thought drilling "a scream." Supervisors "stood opposite their
squads and pretended to know the whole thing. We saluted,
right about faced, etc., with the girls." Her inexperienced
group received extra drilling from Geneva Marsh, a supervisor under Chief Operator Nellie Snow.

But Adele sobered when Marsh placed eleven women under
Adele's command with orders to salute her. Like Grace Banker
playing the Right Honorable Henry Carlton at Barnard College in 1912, the newly graduated co-ed struggled to develop
an authoritative demeanor. Former classmates presented the
greatest difficulty. "Everything is on a strict military basis,"
Adele wrote home. "I am afraid my intimacy with the Seattle
girls must cease, as supervisors are under strict orders to hold
themselves aloof from the common herd." According to Hoppock, it "[is] hard for me to say, 'Do this! Do that!' in true military fashion, but I shall have to come to it."[28]

Hoppock noted cultural differences between women from
other sections of the country. She observed that natives of the
West were not as tractable as women from the East or Mid-
west. As Frederick Jackson Turner wrote in 1890, the frontier
loosened the bonds of civilization. "The Seattle and California
girls are much softer than the others in regard to discipline
and are not used to obeying unquestioningly," Adele remarked
to her mother, "but they have fine material in them"—or so
her chief operator reassured Adele.[29]

Mrs. Berthe Hunt was not one of the California noncon-
formists. During slack hours in the barracks, she read *The
Californiacs,* which she described as "a satire on the boast-
fulness" of her home state. It must have made Hunt's eyes roll:

California, the home of the movie, the Spanish mission, the
golden poppy, the militant labor leader, the turkey trot, the
grizzly-bear, the bunny-hug, progressive politics and most
American slang; California, which at a moment's notice
can produce an earthquake, a volcano, a geyser; California,
where the spring comes in the fall and the fall comes in the
summer and the summer comes in the winter and the
winter never comes at all; California, where everybody is
born beautiful and nobody grows old—that California is
populated mainly with Californiacs.[30]

A quiet, unassuming woman who had run a full household
for her working mother, Berthe Hunt liked order and preci-
sion. When the operators marched raggedly down a Manhattan street after one particularly tiring day, she groaned, "I
do wish the girls were more military & had more snap." Yet
she took pride in them and the impression they made on
New York pedestrians: "We got a lot of clapping though!"
Hunt's main regret was having missed her husband in port.
"Somehow this was a sad day," she wrote soon after arrival.
"Of course I knew R. wasn't here, but still he had been and I
was lonesome."[31]

Drills took place atop the AT&T building at 195 Broadway.
Snow sat on the ground twenty-nine stories below. The operators marched until they were sweaty—then chilled. "We are
beginning to feel miserable," Berthe Hunt wrote in her diary,
"[getting] colds from over heating in drill on that windy roof."
The Signal Corps considered the situation good training for
northern Europe. "The drills included the elementary commands of the school of a soldier," Lieutenant Eugene Hill reported to the chief signal officer in Washington, for which "the

space on the roof of the American Telephone and Telegraph Company's Building was ample and convenient."[32]

Each of the six regiments that served in France was photographed in full regalia atop the AT&T building before sailing. They had survived what one called "a severe mental and physical test."[33] In the first unit, Grace Banker stood in the center of two rows of thirty-two poised, aviation-capped young women. Golden Boy glimmered above them in the fog, clutching his lightning bolts.

The next step was integration into military life with all its routines, including courtesies according to rank. Officers coordinated the movements of large numbers of people in close formation on a tight schedule. Unquestioning politeness, promptness, and respect kept the machine oiled and safe. Rank told each soldier exactly where he or she stood in relation to everyone else and what courtesies were required to whom.

For Signal Corps operators, none of this was clear. Who saluted whom? Should operators rise or remain seated if a superior officer entered the room? And who was a superior officer, considering the operators' relative rank? What about enlisted men?

Such questions became pressing the moment Signal Corps operators entered the regular workforce. They had been drilled in relation to one another but given no instruction on how their rank meshed with the larger gears of protocol. This ambiguity was compounded by etiquette left over from the Victorian era, which restricted casual socializing between men and women. Formal courting preceded the custom of dating, which did not arise until the 1920s. The Signal Corps came into existence on the cusp of an era that the war hastened. Within the space of a few years, women in the United States and England would overturn conventions stretching too far back to date. Hair would go down, hemlines up. Such shocking fashions were epitomized in the 1920 film *The Flapper*. The Hello Girls were too serious and too early to be flappers, but their skirts were still daring enough—cut nine inches above the ground to clear the mud—that the army required black sateen bloomers to preserve operators' modesty against breezes that might lift their hems. The women hated them.[34]

The infamous "Hoboken incident" involving the second unit highlighted the problem, when spontaneous mixing between men and women challenged military procedure and gender-based social codes. Just before sailing in March 1918, a group of bored operators under Inez Crittenden organized a dance for themselves in an empty room at the port of embarkation, partnering with one another. Their laughter and music wafted into the hall, and intrigued men gravitated toward the attractive sound. A government document describes what happened next: "A number of officers, without any introduction, who were looking in at the doorway, came in and suggested that they would join them in the next dance." The men helped the women push tables aside.[35]

Responsible for her unit's reputation, Inez Crittenden was aghast when she peeked into the room. Unwarranted familiarity was a grave breach of propriety. As R. L. Estabrook told Captain Wessen, Mrs. Crittenden immediately whisked "all the operators away" on the presumption that the "young ladies in this service in many cases would not care to meet socially and indiscriminately, every officer."

One pictures Estabrook scratching his head. He asked Wessen, "Has any officer or enlisted man the right to address them?" And, "What should they do, if anything, when encountering an officer or an enlisted man, upon the street?" If either recognized the other, who should salute first?[36]

Estabrook recommended that operators be treated as cadets—below officers but above enlisted men in status. He also forwarded Mrs. Crittenden's proposal: if any officer wished to meet a young lady, he must first consult Mrs. Crittenden's superior officer. That man would then decide whether or not to vouch for the suitor to the chief operator, who would make the introduction to the young lady "at her discretion." The young ladies "may need such protection," Estabrook wrote.[37]

Wessen objected to matchmaking. "The members of this Unit will, it is believed, perform their duties with greater efficiency should they keep their acquaintance among either commissioned or enlisted men at a minimum." He sternly warned that female operators should, as in civilian life, "resent and properly report undue familiarity."

All this was easier said than done. As the captain knew, young men and women would take interest in each another. A female occupation was merging with an all-male workforce. Ernest Wessen was still tussling with army lawyers, whom he hoped would soon militarize the women. He wrote Estabrook that operators were under no "obligation to render military courtesies to superior officers *at the present time*" (italics added). But, Wessen also emphasized, operators should demonstrate "an attitude of respect toward superior officers and enlisted men in the proper performance of their duties."[38]

Wessen was keeping options open. Women were not yet part of the carefully calculated salute-swapping hierarchy, though he was training them for it. He also tacitly agreed that the women's status approached that of a West Point cadet. "There is no occasion in the mind of the undersigned," Wessen wrote, "when an enlisted man should have any right to address them."[39]

For the young women and men affected, these confused policies came down to only one clear principle: no dancing in front of Inez Crittenden until you're out of Hoboken.

The army's parting words concerned sex and drugs. Basic training never ended without homilies on clean living. President Wilson himself exhorted servicemen "to show all men everywhere not only what good soldiers you are, but also what good men you are, keeping yourselves fit and straight in everything, and pure and clean through and through." Berthe Hunt, age thirty-three and married, thought some recruits needed a fuller lecture than the one they got: "D. Mann on sex—fairly good but not very forcible & she took too much for granted."[40]

Some women found such advice a source of amusement, however. Adele Hoppock reported to her sister Eleanor that "Dr. Mann told us that drugs were poison. If anyone tries to make me take any medicine I can remind them of what Dr. Mann said." Merle Egan's friends sent one another into paroxysms of laughter whenever they recalled a YWCA hostess admonishing them to "be sure you wear your brassiere" at army dances in France. "After that," Egan later wrote, "whenever we had a date someone would say, 'Are you wearing your brassiere?'"[41]

The day finally came. The army doctor inspected their throats one last time. The women were ordered to report every half hour if they went anywhere. Then they marched onto the pier and into one of the immense troopships that towered above the docks. With steel sides painted in jumbled angles of beige, blue, and green, the camouflaged vessels bore little resemblance to the passenger liners they had once been. The first unit sailed on the *Celtic*, the world's largest ship at the time it rolled off Belfast's Harland and Wolff quay in 1901.

Every Signal Corps unit shared a sense of awe as they looked out their portholes at the thousands of helmeted men in khaki waiting on the docks, packs on their backs, trench shovels loaded in their kits. Officers closely monitored all the final maneuverings. G2 remained vigilant. The third unit watched in shock as grim officers marched one of its members off the ship at the last moment, suspected of harboring pro-German sympathies.

The last doughboys filed on board. "The embarkation of those men was very impressive—silent, stolid, single file yet with happy faces—in one way it was a sickening sight," Berthe Hunt scribbled in her diary. "Who knows who will return." Three men died that very morning when a ship foundered in New York harbor, the victim of "another German spy in our midst," Hunt thought.[42]

In letters home, Adele Hoppock expressed awareness that war risk insurance was not yet available to army women, but the expectation that a bill in Congress would soon fix the problem. "Provision will be made for us to get it if we want it when it is passed," she reassured her parents. "I am going to take out as much as I can and will put it in Mother's name." A week later, Adele confided to Eleanor, "No insurance can be taken out, they told us today." But Hoppock was undeterred and didn't wish to wait for her sister before hazarding the voyage. "I would rather go ahead now."[43]

As it turned out, a sympathetic House of Representatives passed such a bill two months later, in June 1918, granting insurance to "women serving by official designation with the Army in the American overseas forces as telephone and telegraph operators." The bill went next to the Senate, the more conservative body, where a subcommittee quietly deleted the provision to insure army women. Historian Susan Zeiger writes that despite the vocal consensus on mobilizing women, silent opposition stymied the opening of "military service to them on an equal basis with men."[44]

Few operators felt the moment's gravity as keenly as Chief Operator Grace Banker, who was answerable to the ship's captain for the conduct of his vessel's most unusual passengers and to Captain Ernest Wessen for their success. Banker had double-majored in history and French, which may be why she kept a diary—against regulations—to document an adventure that set a precedent for all American women. "Thirty two girls in my charge, several older than I, youngest nineteen," Banker recorded as the ship sailed on a brackish March morning. The pioneer unit of the female Signal Corps left in a "dismal gray drizzle," the decks quiet and "cleared of all life."[45]

Banker surely knew about the horrific springtime sinking of the *Titanic* en route to New York six years earlier, when she was a college freshman. She may not have known that the *Celtic*—owned by the same company as the *Titanic* and the vessel on which she was sailing—had recently hit a German mine off the Isle of Man, which had killed seventeen. And no one could know that a submarine in the Irish Sea would torpedo though not sink the *Celtic* three weeks after delivering the Hello Girls. If Banker had, it would have made no difference. Like every operator sailing into hazardous waters, she had buttoned her fears under her uniform jacket. "After two months of preparation in the U.S.A., I've crossed the Rubicon now," the student of history wrote. "There can be no turning back."[46]

Miss Grace Banker had signed up for the duration. The question that remained was how near the front Jack Pershing would allow her to get.

6

WILSON ADOPTS SUFFRAGE, AND THE SIGNAL CORPS EMBARKS

WOODROW WILSON was under full sail as well. In October 1917, shortly before Pershing asked for women operators, the president intimated for the very first time that he might support federal action to grant the vote. The war had quickened "the convictions and consciousness of mankind with regard to political questions," he told a delegation of New York suffragists at the White House that fall. "I believe we ought to be quickened to give this question of woman suffrage our immediate consideration."[1]

The suffragists reminded Wilson that while their volunteers were laboring hard to support his efforts on behalf of democracy, grateful leaders elsewhere were enfranchising women. The British House of Commons had passed the Representation of the People Act, and the House of Lords was poised to grant the vote to female property owners and college graduates over the age of thirty. President Wilson acknowledged the contributions of American women, "rendered in abundance," and expressed his opinion that they ought to be similarly rewarded.[2]

Wilson had already moved away from his antisuffrage position and begun to assist in subtle ways. Activists had struggled for several years to convince the Rules Committee of the House of Representatives to establish a special subcommittee to introduce a federal bill. In 1913, the House majority leader, a southern Democrat, quashed the idea when it seemed such a committee might be approved, persuading colleagues that voting was a states rights issue. Four years later, in 1917, Wilson privately convinced the head of the Rules Committee to set up the subcommittee. The president's action eliminated a critical impediment. Soon thereafter, the new House panel on women's suffrage introduced a federal bill.[3]

Wilson's change of heart had been gradual. Because he wrote no private letters explaining it, his motivations remain unclear. Was Wilson sincere or merely using women? Biographer Arthur Link suggests that Wilson's quick remarriage seventeen months after the death of his first wife, Ellen, in 1914 to the vivacious Edith Bolling Galt made Wilson sensitive to the possible negative opinion of women in states where females could already vote. This may have led the political showman to cast his personal vote for a doomed state referendum to enfranchise women in New Jersey in 1915.

Some believe that activists had finally awakened the conscience of the president, who expressed support the next year for state-by-state consideration of the matter. (In the 1916 presidential campaign, both parties added suffrage to their platforms.) These historians focus on the relative efficacy of militant picketers organized by fiery Alice Paul compared with the sedate lobbyists led by Carrie Chapman Catt. Some credit Catt and NAWSA for persuading the president, arguing that abrasive tactics did more harm than good. Others assert that Alice Paul's followers sped the Nineteenth Amendment by eliciting admiration for their courage and cornering Wilson with his own democratic rhetoric.[4]

Beth Behn swings a spotlight on Woodrow Wilson himself to show that his advocacy was more consequential than previously acknowledged. Catt's genteel diplomacy helped the president distinguish between nonthreatening suffragists and less sympathetic activists, whose White House protests spurred charges of disloyalty in wartime. Wilson went far beyond an expedient endorsement in response to NAWSA's pressure. He made the cause his own.[5]

Interactions with bright, dignified suffragists like Carrie Chapman Catt slowly convinced Wilson that a normal, natural, respectable person could be both female and politically engaged. Combined with his PhD in government, these experiences led Wilson to see that democratic government was heading toward fundamental change. "We feel the tide," the president intimated to suffragists in 1916.[6]

From the public's point of view, however, lightning struck when Wilson openly endorsed a federal amendment in January 1918. This was a radical, nearly scandalous step for a southerner to take. It meant risking political capital with Dixie Democrats on a goal so imperative that it was worth using federal power to enfranchise more black voters. Wilson never described the amendment in such terms—doing so would have doomed it—but everyone knew this would be the de facto result in the north and de jure in the south. White suffragists also avoided touting the potential gains for African American women, even shunting them to the back of the 1913 march on Washington to forestall southern reaction. Ida Wells-Barnett, a pioneering journalist and former slave, slipped into the middle of the parade nonetheless.[7] Virulent, unapologetic racism remained a stubborn fact of the political process with which reformers contended, whether or not they shared such prejudices.

In fact, segregation reached new peaks alongside Progressivism. Ironically, southerners were both leading liberals and leading racists. Democratic dominance of Congress during Wilson's early tenure meant Dixie control. The "solid South" was a one-party region that stabilized Democratic control of committee chairmanships that helped Wilson pass key reforms. Southern congressmen, with overwhelming support from their constituencies, proposed and passed the major Progressive legislation mitigating child labor, farm poverty, corporate monopolies, banking instability, and trade restrictions.[8]

The last southern president born before the Civil War, and the first elected in its aftermath, Wilson helped the region regain national influence. This included normalizing Jim Crow. He extended the segregation of public facilities, legalized by the U.S. Supreme Court in *Plessy v. Ferguson* (1898), into federal government. Wilson's administration oversaw the separation of toilets, lunchrooms, and workstations into "colored" and "white." The nation's capitol became a thoroughly southern town once more.

Consequently, in all public speeches, the president never alluded to the most obvious reason why his chief backers opposed a federal amendment on women's suffrage: it enfranchised black women. The president could hardly afford to lose points with those who made his legislative success possible. Nor did he praise the women's movement. Instead, Wilson emphasized two quite new reasons for his change of heart: international opinion and women's wartime service. As Wilson told surprised congressional representatives on the evening of January 9, 1918, "In view of the fact that Great Britain had granted the franchise to women and the general disposition among the Allies to recognize the patriotic services of women in the war against Germany, this country could do no less than follow that example." The Canadian governor-general had informed him over lunch that very day that the wives of soldiers were his best voters. Female suffrage transcended the United States. It was a simple act of "justice to the women of the country and of the world."[9]

Global trends tugged strongly. The first nation with universal male suffrage, America lagged with women. To be a worthy leader of the free world, it must catch up. Progressives had long worried the United States was falling behind. Commonly seen as the vanguard of political progress in the mid-nineteenth century, the United States was considered backward by the start of the twentieth century. In 1912, progressive thinker Walter Weyl wrote in the *New Democracy*, "Today the tables are turned. America no longer teaches democracy to an expectant world."[10] Republican America flickered where it once shined brilliantly.

Wilson's support arose in yet another critical context, which explains why he felt regional obsessions must be set aside. The day before he met with leaders of Congress regarding suffrage, the president gave the most consequential speech of his life. His January 8, 1918, address on war aims, the Fourteen Points, laid the basis for U.S. foreign policy for the next century. In it, Wilson staked out a role for the United States in supporting "a general association of nations" to guarantee the rights of "great and small states alike." Members of the international peace movement had envisioned such a league for decades, but the war pushed the idea to the forefront. Wilson wrote it in the sky.[11]

"The day of conquest and aggrandizement is gone," Wilson asserted optimistically. "What we demand in this war, therefore, is nothing peculiar to ourselves. It is that the world be made fit and safe to live in; and particularly that it be made safe for every peace-loving nation which, like our own, wishes to live its own life, determine its own institutions . . . as against force and selfish aggression."[12]

A second world war revealed Wilson's proclamation as premature, but the president spent the remainder of his career trying to bring it to fruition. Empowering additional voters to support the Fourteen Points and keep Democrats in office became a top presidential priority, especially in the midterm elections slated for fall 1918. Wilson drove his party hard to pass women's suffrage. The midterm elections might spell the fate of the proposed League of Nations.[13]

The question was, Which political party would reap credit for a suffrage amendment? From Wilson's point of view, this was not a matter of mere partisan advantage but a restructuring of the world. Women's suffrage became essential to Wilson's foreign policy. It showed other countries that America was fit to lead. It would also win congressional seats for Democrats who supported Wilson's internationalism.

We cannot know the president's inner feelings and assign exact percentages to his motivations: women's wartime service, the example of other countries, the need for votes to carry out his international agenda, or his desire to appease noisy suffragists. Whatever the combination, the president's personal advocacy had important effects. What the *New York Times* called Wilson's "belated conversion" in January 1918 altered the trajectory of history. The day after he came out on behalf of women, the House of Representatives entertained a prolonged discussion, whose outcome was far from predetermined. The representatives took up the president's arguments with gusto.

Congresswoman Jeannette Rankin of Montana opened the debate and seconded Wilson's point about America's world role: "How can people in other countries who are trying to grasp our plan of democracy avoid stumbling over our logic when we deny the first steps in democracy to our women?" Democratic congressman John Raker of California read off a list of thirteen foreign countries that had granted or were contemplating female suffrage as a reward for wartime service, including rival Germany. Raker quoted British Prime Minister David Lloyd George: "To give the women no voice would be an outrage. . . . That is why the women question has become largely a war question."[14]

Democratic congressman J. Campbell Cantrill of Kentucky acknowledged that a chief argument against female suffrage was that women were unable to bear arms. He had previously opposed the measure, but the war made all manner of unthinkable measures newly reasonable. Cantrill quoted General Pershing as saying, "This war is being fought by women; it is women who suffer and lend courage to us; women are the ones who will deserve honor for their aid in establishing democracy." Wilson's position, combined with Pershing's, had changed Cantrill's mind.[15]

Republican Congressman Philip Campbell of Kansas agreed. "It has been said that women should not be granted the right of suffrage because they cannot bear arms," yet behind the lines in most allied countries they were performing the same work as men in manufacturing arms, munitions, and clothing. The time had come to give women the vote, Campbell argued.[16]

The meeting lasted until 7:00 P.M. Finally, after five hours of debate—and decades of resistance—the House of Representatives approved female suffrage by the precise number of votes required.[17]

Women in the galleries arose en masse to cheer, and a thousand thronged the steps in the chilly dark outside. As the joyful women exited the institution in which they truly belonged for perhaps the first time, one stopped to sing, "Praise God, from whom all blessings flow." Other suffragists took up the song. "From pillar to pillar, the triumphal notes reverberated," Carrie Chapman Catt recalled, and "mounted to the dome of the old Capitol, a sound never heard there before."[18]

Woodrow Wilson did not enfranchise half the adult population of the United States single-handedly. Most of the credit belongs to those who campaigned generation after generation. Nonetheless, as Victoria Brown observes, the president "underwent the most public and important male conversion to woman suffrage in U.S. history." The word "conversion" is noteworthy. The son and grandson of Presbyterian ministers, Wilson had married the daughter of yet another minister. He lived his life in an atmosphere of high mindedness. It is hard to imagine one of America's most ideological presidents promoting a permanent reform of which he did not approve. Women—and international events—had changed his perceptions.[19]

Suffragists were elated. The so-called Susan B. Anthony Amendment had passed one of the two houses of Congress. Carrie Chapman Catt, Alice Paul, and Anna Howard Shaw applauded Wilson for taking his place alongside other world leaders. Paul commented, "All Americans must be proud to have our country join the liberal nations of the world in which women share full liberty with men."[20]

Yet Alice Paul's declaration of a new world order was as premature as Wilson's. The U.S. Senate had yet to vote. It would take much more than Woodrow Wilson's endorsement to change their stubborn minds.

Meanwhile, Signal Corps recruits who were the nation's most unusual embodiment of wartime service were getting their first taste of army life. Their actions would help prove Wilson's point that the world had changed and America must respond.

The crossing to Europe tested everyone's nerves on board. Most of the dangers were gender blind.

"Will anyone ever really know anything again?" Grace Banker wrote in her diary. Like all soldiers pledged to unquestioning obedience, Banker hung on the rumors that circulated constantly regarding submarines, spies, and other ships. She didn't even know whether they were headed to France or England. Such information was above her rank. Her job was to guarantee a creditable performance by America's first women soldiers.

The weather turned bitter on the first unit, which sailed into the North Atlantic in early March. "Even with all my woolen underwear, two coats and a sweater I couldn't stay long on deck," Banker wrote as they departed their first port in Halifax, Canada, where the *Celtic* joined a convoy. Eight or nine vessels sailed within view of one another. German submarines preferred to pick off lone ships—and they were good at it.

During the preceding March, when America still teetered on the edge of belligerency, German submarines had sunk 127 ships on the approach to England, killing 700 men who slipped beneath the waves. The following month, April 1917, submarines demolished 155 ships with a loss of 1,125 lives.

That year, submarines destroyed ships at the rate of five per day.[21]

Another camouflaged troopship, the USS *Leviathan,* flanked Grace Banker's *Celtic.* A group of merchant ships trailed behind. On Banker's other flank, "Dewey's old flagship of Manila days," the USS *Olympia,* steamed protectively. "Must be an awful old tub," Banker wrote. Not that she could see the names of the vessels, which had been painted over. "I don't know where the officers get all this information."[22]

Banker's convoy was lucky to have the old tub. Before the war, the U.S. Navy had spent its dollars from the 1916 Naval Expansion Act on battleships, the largest known as dreadnoughts. Planners assumed that armored vessels would pound one another in epic, blue-water engagements. Alfred Thayer Mahan's unquestioned theory of warfare held that battleships established a nation's standing. They projected invincibility, according to the author of *The Influence of Sea Power on History* (1890).

England and Germany thought so, too, but the kaiser adopted an entirely different strategy when his High Seas Fleet was pinned in port by Britain's Grand Fleet after the disastrous Battle of Jutland in 1916. The dreadnoughts refused to sail out and fight. Germany's submarines carried the burden for the rest of the war, and battleships were rendered practically superfluous. Secretary of the Navy Josephus Daniels realized too late that he should have built maneuverable destroyers to chase submarines rather than stately dreadnoughts to pummel battleships. "O for more destroyers!" Daniels wrote after war was declared. "I wish we could trade the money in dreadnoughts for destroyers already built."[23]

The dearth of escort vessels meant that many ships embarked without them. Berthe Hunt and Adele Hoppock traveled on the *Baltic*, the same troopship on which Pershing had crossed nearly a year earlier. The *Baltic* did not pick up an escort until the fourth day, as it approached "the danger zone" around the British Isles. "What a sight greeted us this morning!" Hunt wrote. "We are right in the midst of a convoy—10 steamers including the man of war—and so close!" She could see the sailors' gray-and-blue uniforms across the water.

Security precautions became tighter as ships neared England. Blasts of the ship's whistle announced emergency drills at unexpected moments, orders came late at night to keep shoes on in bed, lifeboats were lowered. Fast torpedo ships and cigar-shaped minesweepers appeared on the horizon. Convoys began to zigzag. On board the *Baltic*, a soldier was caught signaling and thrown into chains. "Rumors galore," Berthe Hunt wrote. She wondered where her husband was. "Maybe next year we'll be together," she prayed.[24]

Merle Egan's ship, the *Aquitania*—sister ship of the *Lusitania*—was another that steamed out of New York without escort. A few days into the voyage, Egan noticed "an uneasy tension aboard ship." Operators received orders to remain fully uniformed at all times and keep life preservers handy. "We heard our captain was worried," she later wrote. There was no sign of an escort as they entered the most dangerous sector. Daily drills to abandon ship turned solemn. Finally, around noon one day, "a loud cheer went up from all decks," wrote Egan. "Two destroyers began circling our ship. Those tiny boats were mere toys beside the *Aquitania* but we felt much safer."[25]

Soldiers also coped with wild rumors about the war itself. Homesickness and fear plagued many, regardless of gender. There was no indication of victory anytime soon. In the fall of 1918, the War Department was still rushing men to the front and recruiting Signal Corps operators. During October alone, 869 special troop trains transported approximately 320,000 soldiers to military camps and embarkation points in the United States. Pershing promised Allies another three million troops by summer 1919.[26]

Oleda Joure was a Detroit operator who sailed from Hoboken with the sixth unit right after Merle Egan. One of thirteen children, Joure had not anticipated the lonesomeness she would feel among thousands of strangers as they traversed the sea without escort until a day out from England. Like telephone wires, ships transmitted whispers. "It was being circulated that the war would last ten years and I realized that I would be separated from my family for a long time," the Michigan native later recalled. "At night, I used to go up on deck and crawl underneath a lifeboat to cry out loud."[27]

But laughter was as instinctive as tears. Psychological studies reveal that comedy is an innate coping mechanism. In the words of one researcher, "gallows humor is a philosophical attitude, a way to maintain sanity under insane circumstances." Or, more pedantically, "In stressful life-and-death situations, individuals use dark humor as a method of venting their feelings, eliciting social support through the development of group cohesion, and distancing themselves from a situation, ensuring that they can act effectively."[28] Meaning, populations under stress bond over humor.

And so Oleda Joure found herself giggling at another operator's sangfroid when their ship seemed in danger of sinking as it crossed the English Channel. The rocking had grown so violent that many vomited. Michigan operator Corah Bartlett swore "they would have to build a bridge" for her to return to the States. The Signal Corps operators gripped their bunks until the vessel suddenly rose and jerked so dramatically that it appeared to go from full steam ahead to reverse in an instant. Objects flew around the cabin in a deafening clatter.

The woman in the bunk in front of Joure shot into the air and onto the floor as the cabin went black. All Oleda Joure could think was "we are drowning like rats in a hole!" When the lights snapped on a few minutes later, operator Agnes Houley sat on the floor wearing a look of amazement. In Houley's personnel file, an interviewer described the Irish American as "rather a large girl," which Joure confirmed. "A little on the heavy side," Houley landed hard. But the dauntless Bostonian had a way of speaking in the third person that routinely sent bunkmates into peals of laughter. "Aggie T. Houley," she announced, "this is the end, [and] you'll be lucky to make it back to your bunk." If the vessel had sunk right then, Joure said, they would have "gone down laughing, no doubt hysterically."[29]

Rumor spread that their transport had dodged a mine. If the surmise was accurate, passengers were lucky. Two years earlier, a single German mine in the Mediterranean sank the hospital ship HMHS *Britannic*, the largest and last ship in the same class as the ill-fated *Titanic*.

Adrenaline, exhaustion, and a sense of unreality mitigated fear as well. When Grace Banker's ship entered the Irish Sea, greener than any body of water she had ever seen, gossip reported that they had reached the zone of maximum danger. Eight small anti-torpedo boats joined the *Celtic*, "like a lot of skating bugs shooting hither and thither, now slipping under our bow, now suddenly popping up beside us and going through all sorts of weird maneuvers."

As they reached the mouth of the Mersey River outside Liverpool's safe harbor, Banker's immense ship was beached on a sand bar. "What a beautiful target for submarines here in the moonlight," she wrote. Large cruisers tried unsuccessfully to tow the vessel to safety, then abandoned it to escort the merchant ships into harbor. "Can't stay up all night," Banker thought. "Will have to put our trust in these little torpedo boat destroyers which dart so wildly about."[30]

The *Celtic* floated free unharmed later that night, but Adele Hoppock and Berthe Hunt's ship came under direct attack the next month. The first week of the third unit's voyage was unremarkable. Hunt was joyful, thinking it meant her husband was well. "Felt so happy today," she wrote in April 1918, "felt as though my boy was safer than I thought in crossing the ocean & it was such a relief until I heard that if one ship is torpedoed all others run and leave her." Hunt battled loneliness and fear whenever she thought of Reuben.

Rain and rough seas had impeded the *Baltic*'s voyage, and the ship was routed around the top of Ireland to avoid a "super-sub" that was stalking them. As they finally sailed south between Ireland and Scotland, surrounded by large destroyers and small torpedo boats, Adele Hoppock, Berthe Hunt, and other operators witnessed a mile-long chase, when their escorting cruiser and destroyers encountered a submarine. Afterward, the cruiser let out a curtain of smoke that mingled with the fog to screen the giant *Baltic* from predators.

When the women returned to their cabins following lunch, however, they felt a sudden forceful blow. Coolly, but with a sinking feeling, Hunt grabbed a heavy blanket and proceeded on deck, where destroyers were firing on a nearby periscope. Just then, a bomb exploded off the bow, sending a huge geyser of water into the air over the women's heads and spraying their faces. A destroyer's depth charge had hit something a few hundred yards away. Black water boiled to the surface, suggesting damage to a submarine, detectable only from the eruption of oil and air bubbles.

It was impossible to know exactly what had happened. Even after the war, German and British sources frequently conflicted on when, where, and even if submarines had been lost due to surface engagements. The hazards of navigation were considerable in this experimental era. U-boats sometimes just vanished. As a chronicler simply said about one German *Unterseeboot,* "Nothing more was heard from her."[31] Yet the mortal risk of any encounter was undeniable. Troops on board the *Baltic* cheered and danced on deck as they sailed into the mouth of the Mersey. Few transports from the United States were sunk during the war, but everyone knew about the *Lusitania,* in which a single shot killed 1,191. In February 1918, a German torpedo sunk the troopship *Tuscania* in the same Irish Sea, killing 210 doughboys on board. Two weeks after Berthe Hunt and Adele Hoppock's brush in May 1918, yet another British liner carrying American troops tangled with a U-boat. On this occasion, the RMS *Olympic* sliced through the hull of *U-103* and sank it. At the end of that month, German torpedoes off France sent the USS *Lincoln* to the bottom, with 26 deaths.

These events made war real and intensified soldiers' commitment to victory. "Everyone was calm but pretty well frightened," Berthe Hunt wrote on May 6, 1918. Yet she also found the battle "thrilling & I am glad we saw it." Adele Hoppock agreed that the depth charge's aftershock and "spice of danger" kept them motivated.[32]

Passengers on board all transports benefited from the protection of the British Navy. American contributions were important, too. Of the 107 destroyers in the U.S. fleet, 79 entered into convoy duty to reinforce Britain's complement of

400. British cruisers were freed for more aggressive hunting of submarines. The convoy system proved the most effective technique for stemming ship losses, which declined dramatically. The doughboys and Signal Corps women were in much less danger from U-boats than they would have been just a year earlier.[33]

Illness posed a greater risk. Up through the World War I, more soldiers died of disease than battle wounds. (Of 116,516 military deaths during the war, fewer than half were combat fatalities.) "We lost an American boy at sea, Private Brackman of the 7th Engineers," Grace Banker wrote at the end of her first week on board, without stating and perhaps without knowing the cause of death. "We stood at attention while the Captain read the burial service, but the wind carried away his voice and we could not hear a word." Shots echoed over the flag-draped coffin as it slid into the Atlantic at dawn. "Poor boy," she thought. Louise LeBreton, just turned nineteen, and her younger sister Raymonde stood alongside Banker when another soldier was buried at sea later in the crossing, dead from influenza, they thought.[34]

Subsequent units saw worse sights when germs turned virulent that fall, in the second wave of the 1918–1919 flu pandemic. The Spanish influenza had two exceptional characteristics. Unlike most flu viruses, it hit young people hardest. The worst cases also resulted in an unusual bacterial pneumonia that killed swiftly. At Camp Devens outside Boston, sixty-six robust young males died in one day in September 1918. The infection shredded healthy airways. Victims' lungs filled with a blood-tinged froth that poured from the nose and mouth. Dying patients turned blue and even purple from the lack of oxygen, a condition known as cyanosis, then drowned in their own fluids.

The pandemic illness had not yet been spotted when the sixth unit left on the *Olympic*, newly repaired after its tangle with the German U-boat. Called Spanish influenza on the incorrect presumption that it began in Spain, the disease had spread insidiously eastward from Camp Funston, Kansas, then exploded in packed camps and transports during September 1918. On the seventh of that month, one soldier at Camp Devens reported ill. By the eighteenth day, 6,674 had been hospitalized with the worst influenza ever seen in the United States. Five days later, the number rose to 12,604 in a military camp of 45,000 soldiers. Some states suspended the draft.[35]

At sea, quarters were even more crowded. With a status akin to officers, Signal Corps women shared berths with only one or two other operators, but enlisted men slept in bunks sometimes stacked seven tiers high. The biggest ships, like the USS *Leviathan,* carried up to twelve thousand soldiers. Flu hit the transports hard. Of thousands diagnosed, a staggering 6.43 percent died, usually in hospitals after reaching port. According to historian Alfred Crosby, during certain weeks of fall 1918, "the life of a doughboy en route to France was in as much peril as that of a soldier in the Argonne forest," where the AEF was fighting its deadliest offensive.[36]

The USS *Leviathan* was one of the worst affected. An eyewitness said some men "dropped helpless on the dock" prior to sailing in October 1918. Over two thousand developed the flu, worsened by the "intense nervous strain in the war zone and the tremendous rolling of the big ship" in a storm. Every able-bodied person "helped during the great plague," but ninety-nine men died before reaching land. "It was pitiful to see men toppling over dead at your feet," a sailor wrote. "It was like some invisible hand reaching out and suddenly taking them away."[37]

Spanish influenza nearly overwhelmed the *Olympic*, on which the sixth unit sailed in September. When the sick bay grew too crowded, the crew placed mattresses on deck to accommodate the ill and diminish contagion. Michigan operator Oleda Joure, an accomplished singer and pianist, kept up men's spirits by playing a piano someone had hauled on deck. Those with enough strength sang along to the Irving Berlin war tunes she knew by heart.

One song may have been the hit that the Russian immigrant penned when the army drafted him to write patriotic tunes. Irving Berlin's "We're on Our Way to France" received an enthusiastic reception in New York City that summer and fall. "Old Hoboken is bent and broken / From soldiers marching on her pier," the song went. "There's not a minute to spare / . . . For when the Yanks advance / You bet we wanna be there!"

Operator Jane Conroy of Pennsylvania was one of those recruits eager to "be there." When she wrote Ernest Wessen in April 1818, the long-distance operator had already trained bilingual operators for the army, and now she wanted to join them. Conroy had worked for Bell Telephone since age sixteen. If the Signal Corps would take her, she told Wessen, "I would willingly offer my service to go any place at any time." Like Merle Egan, Conroy had to wait until the army could use monolingual operators. Conroy finally took the army oath in August 1918 and shipped the next month.

When the *Olympic* reached England in late September, Jane Conroy left on a stretcher. The twenty-five-year-old was delirious and thrashing. The flu had attacked both lungs. An army hospital admitted Conroy with bronchopneumonia. Her temperature climbed to 105 degrees. Her normal resting pulse of 80 raced to 130 beats per minute. "*Very ill*," the

admitting physician wrote, underlining the words separately for emphasis.[38]

Antiviral drugs and antibiotics did not yet exist. Careful
nursing and luck were the only cures. Jane Conroy's fever
lasted seventeen days. Risking their own lives, nurses placed
warm mustard packs on her chest to dilate the capillaries,
stimulate her nervous system, and help her cough up the
mucus that could drown her. They aspirated her lungs,
sponged her body with alcohol, applied camphorated oil every
hour, gave her salt-solution enemas, and spoon-fed her con-
coctions of milk, eggs, and whiskey. The first week, Conroy
received four hypodermic injections of digitalis to control her
pulse and strengthen her heartbeat.

The entire ship was quarantined for two weeks. When
Oleda Joure finally disembarked, she and other soldiers were
sent to a rest camp sixty miles southwest of London and close
to the port of Southampton. The YMCA ran recreation programs for the waiting soldiers—some anxious to get to the
front, others anxious to avoid it. Silent black-and-white
movies played on a screen under a tent in the fresh air, with
folding chairs for viewers. Word went round that a piano
player was needed, and Oleda Joure's friends pushed her
toward the stage. The twenty-one-year-old soon had everyone
singing tunes like "Over There" and "It's a Long Way to
Tipperary."

When the sixth unit finally repacked for Paris, the YMCA
director approached Joure. Could she stay in England? Camp
Ramsey needed a piano player to entertain the troops. The
Signal Corps operator declined. She had left Michigan to "be a
part of the war effort" and contribute her skills as an operator with four years of switchboard experience. She also had
a brother fighting with the AEF. God willing, they would meet

in France. Nor was Oleda Joure free to resign. Expressing her understanding of her military status—drilled repeatedly into Signal Corps operators—she had been "sworn into the Army." She must serve for the duration, even if it were ten years.[39]

Jane Conroy's fever subsided on her eighteenth day in the hospital. Nurses recorded her having a "very good day." A week later, Conroy "walked a little." General Edgar Russel of the Signal Corps telegrammed London from France on October 15, inquiring after Operator Conroy. The doctor reported that the Pennsylvanian "will recover but will be in the hospital a month more." Conroy arrived for duty in France four weeks after the armistice was signed. She served another eight months, helping to send doughboys home and assisting Woodrow Wilson's peace commission.[40]

Chief Operator Inez Crittenden was not as fortunate. The peppery, propriety-conscious leader of the second unit endured a contentious crossing with her charges. Operators obeyed Crittenden but complained that her orders chaffed. Even the commanding officer on board was dragged into the quarrel. He recommended that Crittenden be reassigned in the "best interest of the service" on account of her "lack of tact" in managing other women.[41] Once in France, however, the San Franciscan proved so efficient at running the Paris telephone exchange that another officer commended her. A few months later, the local office of the U.S. Committee on Public Information petitioned General Edgar Russel to release Crittenden for higher duty, her executive talents were so outstanding. With the secretary of war's official permission, Crittenden transferred to America's public relations bureau in Paris.

Yet the Irish American's service ended tragically. Inez Crittenden died in Paris the day of the armistice. Her burial record does not state why the healthy woman perished at the height of the flu pandemic, but other operators recalled the cause as influenza-related pneumonia. Crittenden's metal coffin in the U.S. military cemetery on Boulevard Washington in Suresnes, France, bore the inscription "Died in the service of her country at Paris (France) November Eleventh, 1918." The rank on her grave read "Chief Operator."[42]

In March 1919, a family member applied for the war risk insurance that Crittenden had purchased a year earlier, witnessed by a second lieutenant in the Signal Corps. A claims officer for the U.S. Treasury ruled that Crittenden was only a civilian and thus her contract was null. No compensation was due.[43]

Men and women shared the hazards of submarines and influenza en route to France, but some of the Signal Corps's challenges were unique, reflecting the trials of integration. Females were curiosities: sometimes resented, sometimes welcomed, always remarkable.

Yet they needed to fit in seamlessly to contribute. To pave the way for others, they had to prove two things. First, they must act and look like soldiers. Second, they had to show that they were not taking away a man's job—which might create resentments too intense to overcome—but doing a vital technical task for which women were peculiarly qualified. They needed to make being citizen-soldiers socially acceptable. Like suffragists, they had to normalize the abnormal.

Grace Banker felt this responsibility keenly. "The days are full for me," she wrote in her diary. The chief operator began each morning on board with roll call, then sick call. She got her unit up on deck for exercises, followed by lifeboat drills and French practice before and after lunch. At nearly every point, the women were under the scrutiny of men who were unsure whether to treat them as soldiers or ladies or both. The women labored under instructions not to fraternize with enlisted men because of their status as quasi-cadets, but also not to speak with officers unless "properly introduced."[44]

The uniforms played a key role in integration, as they symbolized belonging: every soldier answered to the same commands as part of the same army. Uniforms compensated for the ways in which women stood out.

And there were many. During roll call, it was impossible to mask vocal differences. "At first when our names were called we answered 'here' or 'present' but everyone laughed and told us to call out . . . a sort of guttural how like the men do," Banker wrote. Yet she and her unit tried. "We growl as good an imitation as possible but it barely passes and still brings us a laugh." When assembly ended, officers singled them out for special treatment, drawing attention back to their sex: "Dismiss the Ladies and Medical Men" (physicians and corpsmen), someone would call.

Without boot camp training, the women also struggled to perform physical displays for which males had superior preparation. Grace Banker, and probably many onlookers, privately found the operators' attempts at military gymnastics "screamingly funny." One Lieutenant Flugel diligently led them through their paces, but dignity was hard to maintain on a wallowing ship. "When the boat rolls some of the girls lose their balance and sprawl in ungainly manner on the deck," Banker acknowledged, though she was proud of them, too. "They are a fine lot, seldom complaining and for the most part very good sports."[45]

Yet there was no reason why women could not spit and polish as well as men. In this, Banker caught the sensibility of the AEF commander. John Pershing was a stickler who would stop a man in his tracks for an improper salute, insufficient rigor at attention, or unsoldierly dress. "If there was a button unbuttoned, a rifle that was not clean, he would spot it," according to war correspondent Frederick Palmer, who observed Pershing closely. "No top-sergeant could say in more stern reprimand, 'Do you call yourself a soldier?'"[46]

In October 1917, Pershing sent out the order that he intended the entire army, conscripts and all, to adhere to West Point standards: "The rigid attention, the upright bearing, attention to detail, uncomplaining obedience to instruction required of the cadet will be required of every officer and every soldier of our armies in France."[47]

Grace Banker wanted her troops ready for Pershing's cold eye. The chief operator was grateful to find that most honored the uniform as greatly as she did. Only one woman on the *Celtic*, where Signal Corps operators represented all womankind, failed her standards. "Miss G. is the only one who can not seem to look tidy," Banker complained to her diary. "Always has at least one button missing." When the woman showed up one morning with no buttons on one blue sleeve and four on the other, Banker ordered her to cut off two and sew them on the opposite side: "Make yourself as uniform as possible."[48]

Banker herself experienced a panic when she retired one night feeling queasy. "No sooner in my pajamas than there was a firm knock on the door," she recorded. The captain of the ship, a "taciturn old sea dog," wished to see the chief operator. Banker pushed her long hair into her aviator cap and hurriedly buttoned her uniform over "a multitude of sins." Banker later wrote, "Right there I learned the art of camouflage."[49]

In the fifth unit, Merle Egan took seriously the instruction "You are in the Army and must conduct yourself accordingly." This meant immaculate attire. When a fellow operator in Egan's unit lost a button, a superior officer immediately admonished her, "Police your uniform, soldier!" Adele Hoppock wrote her sister Eleanor just before sailing, "A button off our suits would be a catastrophe."[50]

Care paid off. The uniform commanded respect. When Egan sailed out of New York harbor, gazing upon the Statue of Liberty one last time, she encountered a young aviator as she turned from the ship's rail. He stared and then asked, "Why are you on this ship in that uniform?"

Egan looked him straight in the eye. She found her height an advantage when speaking with incredulous men. "Same reason you are," she replied. "I am on my way to France to help win the war."[51] Her answer satisfied him, and they struck up a friendly conversation that continued over the following week. Pilots encountered extraordinary risks, and the aviator, younger than Egan by several years, may have welcomed the expertise of Signal Corps operators. Danger made unusual expedients routine.

Merle Egan's commitment to the uniform that was her license to serve led her to endure black-laced boots so ill fitting they caused injury. Egan's chief operator, Louise Barbour, was a stern taskmaster who had no use for whiners. Unwilling to request special favors, Egan stood through lifeboat drills and endured long marches to mess halls below multiple decks wearing regulation shoes inadequate for her five-foot-eleven frame. When she disembarked in Liverpool, her unit marched to the railway station. Once in Southampton, on the other side of England, even Barbour acknowledged Egan's problem.

An army physician took one look at the swollen toe and prescribed "cold steel." Removing a scalpel from his bag, he lanced it without anesthesia. When Egan calmly thanked him, he looked startled that she hadn't gone into girly hysterics, and Egan was delighted. But she secretly felt like crying when the doctor said she could not continue with her unit. Physical pain troubled her less than the prospect of a travel delay.

The infection was severe, the doctor said. Egan would be assigned to a hospital tent for a week. When another physician later lectured her about her boots, Egan replied, "Yes, sir, I agree they are wrong for me, but I am in the Army and must wear Army shoes." Egan got a better-fitting pair when she rejoined her unit in France a week late.[52]

At the same time that operators were striving to blend in with a man's army, their supervisors had to determine when to stand up and assert their authority vis-à-vis regular military. Once they arrived in England, Grace Banker and others were impressed by the absence of able-bodied males, who seemed to have vanished from the population. Civilian women were carrying the weight of the country. "British women are working everywhere taking the place of men," she observed with a sense of surrealism. The railroad yards swarmed with female laborers. "They wave to us and we wave back," she wrote. For the first unit and others that followed, omnipresent food rationing and brick-like bread confirmed their resolve. The sight of legless and armless veterans heightened it. Able-bodied men reappeared along with WAACs in Southampton, the port of embarkation for France. After a week there, the first unit finally got orders to report for a night crossing to avoid submarine detection.[53]

Awaiting departure in a thick fog, a stewardess working the Channel packet *Normania* warned Grace Banker, "My pal went down last week. We'll probably get it this time." Banker began "to worry about my girls," who had been assigned cabins two decks below. French civilians trying to get home slept on the floor in the narrow passageways. Belgian, British, and French soldiers crowded the stairs, blocked hallways, and milled about the upper decks. Banker went to sleep thinking, "I'll be glad when we are safe in France." The next morning found the ship still in Southampton, trapped by fog. Grace Banker decided then, "Our girls are not going to spend another night in that rat trap below decks."

Banker approached her ally, one Lieutenant Brunelle, who had been appointed to accompany the unit. Together they cornered the manager who had barred American women from the dining room—the only place serving food—and the first-class berths on the top deck. Banker and her colleague "had a row with the British official in charge," demanding access to the dining room and any unoccupied staterooms. The suspicious seaman, a veteran of the London–Africa–America route, rejected their requests. Banker persisted. After "a long conference," the captain relented. "We won," Banker wrote triumphantly, "for we got all the girls up from below decks."

The skirmish proved worthwhile. The sun shone weakly when they finally slipped out of Portsmouth that afternoon, and fog descended as night fell. Faint whistles came across the water from other ships. Banker wondered if submarines lurking near the surface could hear them. "Some one started to sing, but then next moment a sailor forbade it," Banker wrote. "Nothing carries like music over the water." She couldn't see more than a foot in front of her. Her charges huddled around a smokestack for warmth. Banker and Lieutenant Brunelle pointed out the lifeboat where they should rendezvous if attacked. "There was a tenseness in the atmosphere."[54]

Banker and operator Susanne Prevot, another New Yorker with a keen itch to see action, leaned against one another on a bench sometime after midnight. Others went below to their new berths. Lieutenant Brunelle paced, trying to see in the soupy miasma. Banker had barely started to doze when, to her horror, the ship lurched and cut its engines. Her eyes blinked open. Phosphorescent flares leaped skyward, and brownish posts bobbed up from the brine. The net of a submarine trap surrounded them. The shadow of a large vessel immediately loomed in the dim, bearing down to ram. Within seconds, the destroyer reversed its engines, bumping the packet only lightly. The French patrol realized just in time that it had caught a guppy, not a shark.

Banker and Prevot went below an hour later, too cold to stand watch any longer. They tumbled into a single bunk with two other Signal Corps operators, one of whom ended up on the floor. The four women awoke to the sound of French cowbells. They had but one more leg of the journey to complete before taking up their stations.

The United States had been in the war nearly a year when the first three units arrived in March and April 1918. Strangely, the operators were not even late. American soldiers were just amassing and had not fought a single battle of their own—of which their bleeding partners were painfully aware.

U.S. policy since George Washington had been to avoid "interweaving our destiny with that of any part of Europe."

The first president warned in 1796 that "permanent alliances" tended to yield "overgrown military establishments" that would "entangle our peace and prosperity in the toils of European ambition." Breaking this tradition while trying to uphold it, President Wilson decided the United States would be an "Associated Power," not an ally. He and Pershing wanted a separate American army, not hastily mustered doughboys filling holes in decimated French and British regiments that lost men as quickly as they were replaced.

Allied strategies had produced a war of attrition. Pershing wanted to strike only when fully ready. Black Jack counted on the psychological advantage to be gained by a powerful knockout blow. Rather than dribbling troops onto the front as they arrived, he would hold them back until he had a million freshly trained, well-rested men for a crushing offensive to bust apart German defenses.[55] Pershing gambled that France and Britain could hold the Germans until then. Of course, the strategy was easier to contemplate if one's own capital was not under the gun.

This policy also delayed the U.S. war effort. The AEF spent a year laying the groundwork in France before it made any telling contribution. Historian John Terraine observes that for the Allies, it was a "puzzling and disappointing fact that although the United States had been in the war since April 1917, by December [1917] there were only four American divisions (about 140,000 men) in France, of which only one was actually in the line, on a quiet front."[56]

The first three units, including Grace Banker's, arrived at a frightening low point. Vladimir Lenin, head of Russia's new Bolshevik government, had approved an ignominious but expedient peace with Germany three weeks earlier. The Treaty of Brest-Litovsk exchanged the prosperous western territories of the former Romanov Empire for a cessation of hostilities. Russia lost roughly 90 percent of its coal deposits, 50 percent of its heavy industry, and 30 percent of its population, but the Communists could concentrate on tightening their hold. Newly enriched, and unencumbered by the eastern front, Germany turned its undivided attention to the western front, where 300,000 well-fed, newly equipped Americans were arriving every month.[57]

Now was their moment, the kaiser's commanders thought. The Russians were out, the Americans not yet fully in, and the British and French stretched past their limits. On March 21, two days before Louise and Raymonde LeBreton returned to the land of their birth, the kaiser's army launched a stunning spring offensive. Supported by a massive artillery barrage, a million men attacked. Once again, German armed forces came within shelling distance of Paris and victory.[58]

The assault began with drenching gas attacks and a five-hour hurricane bombardment. Waves of Germans overran their stunned enemies. In a single day on the Somme River, 7,500 British soldiers were killed, 10,000 wounded, and 21,000 taken prisoner. England lost more artillery in twenty-four hours than it had in any other attack. It was the second worst day of the war for Britain (which suffered 19,000 deaths on July 1, 1916). The Fifth Army of the BEF was shattered. British commander Douglas Haig admitted that his men had their "backs to the wall."[59]

The French fought ferociously while the Germans advanced toward the Marne River for the second time. Outside Paris, Germany set up "Big Bertha," a forty-seven-ton mobile cannon nicknamed for the heiress of Krupp, the arms manufacturer. Bertha fired shells into the heart of the ancient capital. French general Petain confided to Prime Minister

Georges Clemenceau that he believed England would surrender within fifteen days. France might follow.[60]

In eight days, German troops had advanced forty miles, taken 70,000 prisoners, and inflicted 200,000 casualties.

Signal Corps Captain William Vivian met the first unit of Signal Corps women at the port of Le Havre that same week. When they boarded a train for Paris, he told them their destination was under bombardment. The women were undeterred. Instead, Berthe Hunt felt inspired by his trust in their commitment and competence. "It certainly is flattering to see how proud these Signal Corps officers seem to be of the girls," she wrote. They pressed on.[61]

Grace Banker observed that English WAACs were "everywhere," loading heavy trunks onto French railway cars. Before Banker left the port of Southampton, one Dr. Birnie warned her, "The W.A.A.C.S, God help them!" He told Banker, "England doesn't know how to take care of her girls." The picture he painted led her to wonder what consequences women faced for their well-intentioned service. "It may be a good thing to have one's eyes opened," she wrote cryptically. Did Birnie tell Banker that English women in uniform were not even entitled to a salute?

Later operators observed the treatment of WAACs with concern as well. Ellen Turner, who sailed to Europe with the fourth unit, noticed that WAACs stood in the "broiling sun" while Signal Corps operators enjoyed the shade of an awning and privileges of commissioned soldiers. WAACs were "not allowed to associate with officers," Turner told her parents, "and I think it's pretty hard on them, too, for some of them I imagine, are very nice girls."[62]

In France, Britain's uniformed women, many from the working classes, operated the docks and sped soldiers on their journeys. "It gives me a funny feeling to see them doing a man's work," Grace Banker admitted, expressing a common apprehension.[63] Could women step into such roles and remain "normal, natural" females?

The chief operator was unaware that U.S. Secretary of War Newton Baker felt much the same about her. Nonetheless, Banker, her colleagues, and the American government took comfort in the notion that—as extraordinary as it was to see women in the uniform and insignia of the U.S. Army—at least they were doing women's work, and their cadet-like status admitted them to military gentility. But the central paradox remained intact: their job was crucial to modern warfare, yet the War Department swore to keep the army a male preserve.

Like many subsequent Signal Corps operators, the first unit arrived in Paris long after dark. Streetlights were switched off to avoid illuminating targets for German aircraft. Dim blue lamps guided the women as they toted their gear past statues, churches, and buildings reinforced with sandbags against bomb attacks. "We could hardly see the curbstones," Louise LeBreton later recalled. Parisians who passed them in the dark mistook them for refugees clinging to their last possessions.[64]

Their destination was the YMCA hostel, Hotel Petrograd, where they finally had a chance to rest. After less than an hour's sleep in their temporary quarters, however, they were awakened by an unearthly howl that soon became familiar, depending on where they were stationed. The hoarse shriek of the air raid siren pulled Grace Banker, the LeBreton sisters, and twenty-nine other women from their beds. Hotel personnel ran through the corridors shouting "Alerte! Alerte!"

Unused to physical danger, some underestimated the threat. Cordelia Dupuis, the North Dakotan whom an AT&T

interviewer described as possessing an "absolute absence of nerves," was tired and wanted to sleep, "but the French maids made us go to the cellar, five flights down. We slid down the banisters in our nightgowns," the twenty-one-year-old recalled. "But we were plenty scared the next day when we saw that huge hole next to our hotel."[65]

Banker counted her operators when they reached the bomb shelter. "Two girls missing," she found, and struggled her way back up the dark stairs. By the time Banker found her errant charges, the horrible siren that sounded like "some weird monster in distress" had stopped.[66]

Managers of the hostel noted the courage and calm of the newly arrived women. They had evacuated in an orderly fashion, then "sat in the little room in the dark listening to the bombardment, quite composed, and were up the next morning at 6 o'clock with not a murmur." The top ranking official for the YWCA, Henrietta Roelofs, reported that she knew then that "these girls were going to make good." A Signal Corps captain told his chief that although explosives fell close by, "During the time the city was shelled with long range guns and by air planes at night, the operators conducted themselves in a very commendable manner."[67]

A Signal Corps photographer snapped their picture for posterity the next day, and the first thirty-three women soldiers received their assignments: one group staying in Paris; another going to Tours; a third heading for General Pershing's headquarters in Chaumont, a city of fifteen thousand inhabitants located 150 miles southeast of Paris. Newspapers across the United States and proud publications of ATT reproduced the photo.

It was a war that trapped generals in their headquarters, to which all telephone cables led. "A vast, intricate net of cables linked the trenches to the command posts all along the fronts," Terraine notes. "Generals had to be at the nodal points of this net—the ONLY place where it was possible to know what was going on at the end of a wire."[68]

Grace Banker and Louise LeBreton counted themselves lucky to be joining Pershing. "We are going to the bombing section," Banker wrote happily. Her group shared the crowded train to Chaumont with French soldiers and civilians, and a group of "red faced YMCA men" who declined to relinquish a seat to the American women in uniform. The operators hoped to start upon arrival, but a "very gruff Signal Corps lieutenant" snubbed them, and told them they would have to wait. "I don't think he likes women," Grace Banker wrote of Lieutenant J. W. Riser. She contented herself with awaiting orders.

Finally, the next day, "The girls took their places at the boards," thrilled to connect the army's lifeline. The hours flew by. Soldiers and officers expressed astonishment when they heard the voices of American women and the familiar AT&T greeting so very far from home. In her diary, Grace Banker recorded the reaction of one man in particular: "When you said 'Number please,' I couldn't answer," the soldier told the operator; "there was a lump in my throat."

The Signal Corps operators had shown their mettle by crossing what newspapers called the "submarine-infested" sea. Now they had to demonstrate their utility to the U.S. Army in the first global crisis of the twentieth century.

7

AMERICANS FIND THEIR WAY, OVER THERE

L IKE SIGNAL CORPS operators, the United States had
    much to prove. Could Uncle Sam be a good soldier?
Would he show up in sufficient force, in sufficient time, with
sufficient nerve? Would the nation stay until it was "over, over
there?"

Allies had doubts. The United States fought its first engagement thirteen months after declaring war and nearly
four years after the assassination of Archduke Franz Ferdinand. America could not win the war for Allies who had already defended every inch of turf at unimaginable cost. But
the U.S. Army, if sufficiently large and determined, could
force Germany to rethink how much longer to go on. To help
end the bloodletting, America had to credibly show it added
enough weight to the scale to make further German resistance futile. John Pershing endured the Allies' scorn and re-
sentment, and withstood extraordinary pressure to give them
his men in order to construct the strongest army possible.

Britain lost two times more soldiers in the Great War than
in the next. France suffered six times more deaths in World
War I than in World War II. Of the eight million Frenchmen
mobilized from 1914 to 1918 to expel the German invasion,

73 percent were wounded, of whom 1,357,800 died. Half of all men between age eighteen and twenty-eight were killed or maimed. Russian and German losses were even greater. In France, the war leveled forests, fields, villages, mines, steel plants, and factories across the north. Eight million acres were chewed into wasteland.[1]

In comparison, 7 percent of American troops were wounded or fell ill. Of these, 116,516 died. Thanks to the width of the Atlantic Ocean, not a single American home or farm was destroyed. Nonetheless, American resolve in the face of unprecedented violence would determine the outcome of the conflict as it entered its fifth year.

Industrial progress created the daunting odds that U.S. troops faced in 1918—and that had stymied either side from winning up till then, turning Germany's anticipated six-week campaign into a numbing nightmare. Factories of the early twentieth century made munitions of frightening power available in staggering quantities. During the Franco-Prussian War of 1870, a cannon averaged forty rounds a month. By 1918, French and German cannons fired 450 rounds per day. Machine guns, invented only a few decades earlier, fired 600 bullets per minute with a range of more than 1,000 yards.[2]

A storm of steel met men whose commanders led them "over the top" in suicidal attempts to advance the line. Robert Graves, a British poet and soldier, compared the western front with a "Sausage Machine, because it was fed with live men, churned out corpses, and remained firmly screwed in place."[3]

Even humbler technologies played a role. American ranchers in the 1870s adopted a form of fencing known as barbed wire to curtail the movement of cattle on the Great Plains and vast expanses of Texas. By the turn of the century, so much wire crisscrossed the West that demand for cowboys to corral strays dried up. In the Great War, Europeans coiled barbed wire tens of yards deep in front of their trenches, making no man's land virtually impassable. Machine gun bullets passed through the vicious tangles without destroying them. Only armored vehicles, or prolonged bombardment, could do so.[4]

For much of the war, Europeans had just enough technology to pin one another down but not enough to break through the lines. In 1914, the belligerents had only five hundred planes among them. By the end of the war, Britain alone had fifty thousand. (The U.S. Army possessed only thirty-five trained fliers in 1917; by war's end, it had cobbled a force of twenty-seven hundred planes.) More important was France and Britain's deployment of tanks in 1917 and 1918, allowing crews to penetrate the deadly fire zone, crush barbed-wire defenses, and blaze a route for infantry running behind. Motorized assault vehicles arrived too late to prevent years of slaughter, though they proved critical in closing battles.[5]

Both sides had finally begun to break out of the trenches by the time the Hello Girls arrived. The German spring offensive marked a change in tactics. *Sturmtruppen* (stormtrooper assault units) carrying lightweight machine guns, mortars, grenades, and flamethrowers ran through gaps created by intense bombardment of British lines to capture points behind them, with the hope of cutting the English off from the French.

The Germans advanced so rapidly through these breaches in March and April 1918 that they outran their lines of supply and had to fall back to more defensible redoubts. Nonetheless, they captured British and French positions previously won at the cost of hundreds of thousands of lives. Paris braced once more for invasion. Up to a million civilians fled.

British forces gathered northwest of Paris near the coast-line, close to their escape route. French forces ranged outward from Paris itself, in the middle of the country. Together they held off repeated German attacks, hoping their new partner would respond before all was lost.

In January 1918, the French General Staff secretly assessed the American Expeditionary Forces as a "weak asset."[6] Woodrow Wilson's critics agreed. The United States had been reluctant to get into the war, and once in, reluctant to attack. To confirm its worth as an ally, America had to finish building its army and send troops forward. To confirm their worth as soldiers and citizens, Signal Corps operators needed to hit the ground running.

The first one hundred Hello Girls arrived well before most American troops.[7] This was part of Pershing's careful mobilization plan, which seemed infernally slow and even irresponsible to French and British commanders. The U.S. Army was still assembling and training in Lorraine, southeast of Paris and near the German border, and did not reach the one million mark until July 1918. Lorraine was a relatively quiet sector. Men climbed in and out of pretend trenches, learned to withstand gas attacks, and practiced shooting. When the Hello Girls arrived, there were only 287,000 Americans in France, with no full divisions near the battle zone.[8]

Scattered units rotated through French and British forces to give them experience. A few Americans died. The U.S. Army's First Infantry Division left Lorraine in April to help the French First Army stay the German offensive. In early May, Signal Corps units serving with European troops absorbed gas attacks while reconnecting broken telephone wires. As Sergeant Howard Cooper reported, he and his men worked sixteen hours with their masks on, "repairing line after line, and [while] man after man kept falling to the ground," their skin burned black from mustard gas.[9]

Most American troops were not yet in danger, though. The Signal Corps was still building its communications infrastructure across France as military preparations went feverishly forward. They eventually erected 1,724 miles of permanent telephone lines from LeHavre to Marseille, strung with nearly 21,000 miles of wire. A station in London received cross-Channel communications. Male soldiers operated combat switchboards and the smaller exchanges, of which there were approximately two hundred. By the end of the war, Signal Corps circuits served 8,152 "subscribers," many in muddy trenches.[10]

Like other new troops, Signal Corps women faced four initial tasks. They needed to master their assignments, develop effective relationships with coworkers, settle into military life, and establish a support system to endure looming challenges. And they had to accomplish these prosaic tasks in a setting that reminded them constantly of their mission's urgency: maimed young men, roadside graves, devastated villages, and short-tempered superiors under horrific strain.

Mastering their assignment was both the easiest and the hardest challenge. Military exchanges functioned much like civilian ones, though calls were considerably more important. As the orientation memo Merle Egan received in August 1918 told operators, the army would accept nothing less than excellence: "The results of battles and lives of thousands of our soldiers frequently depend upon the speed and accuracy with which telephone operators complete connections."[11]

Most women soon became practiced at the job but found the workload and responsibility daunting. Male soldiers trained four to six months before shipping to France for further instruction. Signal Corps operators often trained less than a month and began work upon arrival. Following Berthe Hunt's first Monday, when she walked past trenches, air-raid shelters, German POWs, and troop trains filled with infantry on her way to the exchange, she wrote, "Oh, what a day! So busy, so mixed, & we so green! I bet we each had a little cry."[12]

Seven days later, the boom of enemy artillery shook windows and walls so badly that "the phones were almost impossible to use at times," Hunt reported in her diary. A few weeks later, when Paris was under continuous attack and Signal Corps women drilled with leaky gas masks, she penned, "This has been a hard day—work, work, work & I was unstrung and nearly lost my self-control." Grace Banker fell into bed exhausted most nights. "I just managed, managed, managed all day until I was all in."[13]

No one could tell them to take it easier. The Signal Corps needed miracles. When General Pershing sent his original cablegram, the switchboard at his Chaumont headquarters could handle only eighteen hundred local calls per day and two hundred long-distance calls. The entire U.S. Army in France was making only four thousand local calls a day, a number grossly inadequate to coordinate a force of any size.

The army had to establish in months what had taken AT&T decades to achieve: universal service across a nation. This required not only a solid infrastructure, which the army was steadily building, but also personnel using the best procedures.

Signal Corps records reveal an organizational model based on scientific management, sometimes called Taylorism for its inventor Frederick Taylor. Taylor had been poised to enter Harvard in the 1870s when "eye strain" gave him an excuse to enter a manual occupation he found more interesting: working in a machine shop. Taylor inherited his drive from his mother—an ardent abolitionist, feminist, and friend of suffragist Lucretia Mott.[14] The idea of Taylorism was to standardize every phase in any production process. In wide use after the 1890s, scientific management achieved with human beings what Eli Whitney achieved earlier with machines: turning people into interchangeable parts. Taylorism traded individual style for uniform quality and speed. If an operator used the same abbreviated protocol for every call, communication was compressed. Connections sped up. In peacetime settings, Taylorism made telephone work arguably the most tightly supervised female occupation. Unions called it oppressive.[15]

Taylorism proved good for war, however. Speed saved lives. The principles of scientific management fit well with military discipline, which also minimized choice. Officers wanted soldiers to do precisely what they were told, in the exact ways specified. Edgar Russel's team issued a handbook to arriving Signal Corps women. The *Military Telephone Regulations* mimicked domestic procedures designed to soothe customers and process calls as fast as possible. The handbook dictated every aspect of telephone work, from the timbre of voice to the pronunciation of numbers, along with supplemental wartime rules.

As the handbook reminded recruits, subscribers could not see an operator's face, so "impressions are formed chiefly from the manner in which words are spoken." Switchboard soldiers must cultivate "a distinct, clear and cheerful tone of voice." The handbook prescribed rote phrases for every occasion.

Operators must project caring while acting like machines. Calls should begin, *"Number, please?"*

Operators must then repeat the number, speaking "each part of the number separately with a rising inflection." Or, if a caller's voice was indistinct, *"I beg your pardon?"* When the operator couldn't put a call through, she was to say, *"The line is busy."* If communications had been blown up, she should reply, *"The line is out of order."* Under no circumstances should an operator reveal her name or location. The former was irrelevant, the latter secret.[16]

Gender and cultural dynamics complicated the rules. Operators constantly balanced the demands of courtesy with the dictates of efficiency. Intrigued male callers asked for operators' names or tried to prolong the conversation. As Berthe Hunt wrote two months into her service in France, "same old story of surprised American soldier." When the caller found she was American, he shouted out to someone nearby, "Come Jim, here is a female woman!" Once both soldiers had ears to the receiver they made her repeat twice, "I am a sure to goodness American." One curious soldier who could not coax an operator to reveal her name wrote the exchange asking for a formal introduction to the operator with the "smile in her voice."[17]

The precise, persnickety requirements reflected a larger purpose. As a war correspondent for the *Los Angeles Times* wrote when Paris was being bombed in May 1918, "I reckon that the well-modulated, courteous and very American accents of a hello girl dripping in at the left ear have much the same effect on a homesick reporter that the soothing hand of a nurse has on a sick soldier." Harry Williams expressed the sentiments of many men exposed to war's disorienting hazards when he observed that the consistent service patterns of the Signal Corps reassured him that all would be well: "The whole thing sounded so much like Los Angeles that I could have given three cheers for the Bell telephone system, while I breathed a silent prayer that all the hello girls in the world might prosper and marry well."[18]

Cultural differences occasioned more frustrating delays. Toll calls often went through French circuits, especially when U.S. lines went down, requiring help from a government-run system untouched by scientific management. French operators expected what Signal Corps personnel considered excessive courtesies, from chitchat about the weather to honeyed inquiries after one another's health, before proceeding to business. Snappy American procedures violated French norms. American operators came across as rude, and the French could be rude in reply.

Berthe Hunt, with family in Paris, pronounced her counterparts "nice bright French girls." But even she found French Central—or "Mademoiselle," as Signal Corps operators called their counterparts—not "very kind" in the face of urgent requests. Hunt's hardest days were those she staffed the toll board that connected to French operators, whom she described in her diary as "very hard to work with and have no patience."[19]

Adele Hoppock also experienced her counterparts as "temperamental and exasperating," and found herself occasionally driven to despair when a local operator simply dropped a call without explanation. Mademoiselle "disrupted at the slightest whim a laboriously built connection, often times even dispensing with the ominous 'Je coupe!' ('I cut you off!')."[20]

Grace Banker found that French operators had "an easy, unhurried philosophy of life, perhaps to be admired, except in times of stress." She learned to assign "my most wheedling operator on the French board" and lie about things like the weather. Local operators refused to answer electrical switch-boards during thunderstorms, deeming it "too dangerous," so Banker told her operators not to mention signs of oncoming rain.[21]

French Central complained about the frequency of Signal Corps calls as well. "You are unbearable, you ring too much, it gets on my nerves," they sometimes told the American women. If Frenchwomen thought Americans were camping too long on their circuits, they simply ended the call. A con-nection requested at ten in the morning might not be granted until two in the afternoon. Yet Banker's experience of Gallic soldiers was different. When her operators interfaced with French officers near the front, "they gave us the towns quickly," Banker later recalled. "Perhaps they knew only too well the difference a few minutes might make."[22]

The *Military Telephone Regulations* handbook reveals one source of the culture clash. American rules prohibited discon-necting any call before subscribers stopped talking—unless General Pershing himself was on the line. Operators had to be brief, not subscribers. And their actions were carefully timed. Rings, for example, were to last two seconds. If no one answered, operators were to "continue to ring about every ten seconds for a period of ninety seconds." Common battery and magneto boards required different, equally precise, maneu-vers and phrases. Unlike French operators, American women coped with highly regimented procedures.

In addition to rules calculated to increase speed, the hand-book contained general instructions. Operators were to practice teamwork, maintain their uniforms, aspire to "the best possible physical condition," keep living quarters that would pass "military inspection at any time," and conduct themselves so as to "win the praise and admiration of not only [their] associates but of the French people as well."[23]

Above all, Signal Corps operators were to practice discretion. "All communications will be treated with the utmost secrecy. . . . Upon an operator's faithfulness to this trust may rest the lives of thousands of men." Prosecution faced anyone who failed to comply. "It is a court martial offense to give anyone, except the proper authorities through military channels, any information whatever regarding communications." Grace Banker summarized army instructions in plainer terms: "We are told to keep our mouths shut, ask no questions and never discuss anything."[24]

Superior officers warned operators to be mindful of enemy spying. Germans were sometimes close enough to intercept a ground wire. "We have to be careful too of listening in on the lines," Berthe Hunt noted; "it makes a dull noise, they say, when they are tapped."[25]

To disguise army troop movements, the Signal Corps changed code names frequently, requiring operators to memorize new settings. One week, callers might ask for Montana, Buster, Bonehead, or Podunk; another week, Waterfall, Wabash, Wilson, or Willow. The mad rush sounded like a scene out of *Alice in Wonderland*, Grace Banker thought. "I heard one of the girls say desperately, 'Can't I get Uncle?' and another, 'No, I didn't get Jam.'"[26]

Their dedication paid off. Army telephone service tripled. The chief signal officer reported to Congress that local connections increased from thirteen thousand calls per day in January to thirty-six thousand calls per day in July, following the arrival of "specially trained personnel"—meaning women. Experienced supervisors like Grace Banker, Inez Crittenden,

Merle Egan, and Louise Barbour monitored the "accuracy, the speed of connection, and the uniformity in method of handling calls" to deliver the crisp service that AT&T's female operating force had made famous back home. The expansion of the army's infrastructure contributed, but women's labor was critical.[27]

And it kept getting better. Under the watchful eye of female supervisors, defects were noted and improvements demanded. The operating force actually achieved faster connection times during wartime than in large cities back home. In July 1918, when mobilization approached new peaks and additional Signal Corps units arrived, toll calls quintupled. By November 11, the Signal Corps was averaging 4,000 toll and 150,000 local calls every day. By war's end, they had connected more than 26 million calls.[28]

The women's efficiency, and telephone technology itself, transformed military expectations. Other forms of signaling remained in use—from radios and lamps to pigeons and flags—but commanders used them sparingly, as backups. "The American has learned to think in terms of the telephone," Chief Signal Officer George Squier reported, "and so it was natural for a brigade commander, whose telephone lines had been shot out, to send this radio message to his division: 'I am absolutely out of all communication.'"[29]

Excellent job performance helped the women with their second task: developing good relationships with skeptical men.

Lt. J. W. Riser admitted to Grace Banker "he was mad when he first heard he was to have charge of a lot of 'female operators.'" It explained his grumpy reception the day they arrived. But they converted him within the week. "Now he is very proud of us all," Banker wrote shortly after taking charge at Chaumont. "Every time any one says anything nice about

us, he promptly reports in a most beaming manner." The lieutenant seemed particularly tickled, she observed, by a fellow
officer's compliment that his subordinates "had lots of pep and
were regular soldiers."[30]

Merle Egan encountered a more challenging problem when
she was placed over men, an assignment that challenged
gender norms more directly but revealed once again the AEF's
willingness to do anything necessary to get the job done.
When Egan arrived in Tours, France, her training skills became quickly apparent. Captain Edward Stannard placed
Egan in charge of teaching men on their way to the front to
operate magneto switchboards, equipment common in the
Montana backwoods. As the war amped up in September 1918,
Egan trained successive waves of soldiers, compressing a two-
week training program into three days.

Egan's students sometimes balked at what they considered
a feminine occupation. They asked, "If I'm going to be an operator, where are my skirts?" Egan replied that any man could
carry a gun, but "the safety of a whole division might depend
on the switchboard one of them was operating." One particularly ornery sergeant refused instruction altogether. "I am not
going to report to any woman," he told her. Egan's supervisor
gave her leave to "assign him to K.P.," kitchen patrol, the
lowest form of punishment.

"I was treated to all the cuss words in his vocabulary," Egan
recalled, but she stood her ground. The man served his sentence in the kitchen and came back one of her best students.
"We parted friends."[31]

Signal Corps personnel also won admirers by showing a
sense of humor that endeared them to men who must have
worried that women might be prissy and reserved. On duty,
Grace Banker was rigorously professional. Off duty, she played

practical jokes, climbed through windows to avoid boring visitors, gallantly welcomed officers who wanted to see "real American girls," and roared with laughter when fast rides in the back of bumpy trucks required hanging on for dear life.

Like many operators, Banker showed a useful knack for befriending men—and respect for the greater sacrifices some had made. One man, a Lieutenant Bradford, who "kept [the girls] laughing" with his funny songs and jokes, couldn't take the chair Banker offered, as he was "dreadfully wounded in his back and legs." Another gave her his left hand to shake, having lost his right. Like the best officers, Banker developed the skill of conveying compassion without pity. To men faced with a young woman, this must have been particularly welcome.[32]

Berthe Hunt, mature and married, proved equally adept at establishing rapport. The "boys" of the Signal Corps gave her a cat when she was transferred closer to the front. "Camouflage is its name & the name fits," she wrote. For her part, Louise LeBreton refused to accept defeat when an angry young officer made clear his displeasure at working alongside women. LeBreton began "a campaign of my own to show him that we were not so bad, that we were soldiers, too." Louise bested her adversary so thoroughly that they married after the war.[33]

Banker made a firm ally of Lieutenant Riser the day she invited him to join the operators for dinner at the Signal Corps residence. The lieutenant was extremely tardy. Grace said nothing, but she and Marie Gagnon of North Dakota set off an alarm clock under his chair, "to let him see how late it was." Riser jumped—then laughed. "Fortunately he took it as a good joke," Banker wrote. The chief operator had treated a male officer as a social equal, and it liberated him to act as one.

Lieutenant Riser consulted Grace Banker frequently after that evening, and her diary afterward was filled with references to their collaboration. She played along when Riser made out that he was "a fiendish task master" in front of other men ("much bluff," she noted), and protected him when rear echelon officers whom Riser despised pumped her for information on his unit. A comradely atmosphere blossomed. When Brigadier General Edgar Russel and Colonel Parker Hitt inspected switchboard operations at AEF headquarters the month before the first American battle, "they complimented me on my work here," Banker noted.[34]

Support at the highest levels of the AEF—if not the bureaucratic War Department—also eased integration. General Pershing himself reached out to female troops. Women mustered and stood alongside men at Chaumont in the drenching rain or scorching sun whenever Pershing held a general review. They saluted him, and he saluted back. On one occasion, Pershing personally inspected their Chaumont living quarters. The women rushed their dinners and hurried into clean uniforms to greet the general, who arrived on short notice at 7:30 P.M., accompanied by other top officers.

The AEF commander in chief hardly needed to inspect the billet of any particular group of soldiers, especially ones of such low or at least indeterminate rank. Doing so sent a message to fellow officers, who paid close attention. Pershing told Grace Banker with obvious amusement that when his military household heard he was going to examine the women's billet, "they all insisted on coming along."

The intense, handsome, fifty-seven-year-old Pershing spoke with each operator upon arrival, shining his starry fame on every individual. Grace showed him their quarters, including even the kitchen, where an ecstatic French cook stood at rigid attention. "Oh, I always want to see the kitchen," Pershing insisted when Banker started to steer him past it, as beneath notice. The commander complimented the chef on the tidiness of her domain. "I'll say his inspections are thorough," Banker noted.

When Pershing pronounced himself satisfied, Banker demurred that their billet hardly qualified as "soldiering," as they were too comfortable, especially when compared with the general's encampment two years earlier on the Mexican border. Pershing smiled at the self-effacing suggestion that she had followed his military career and was ready for tougher challenges. The accommodations at Chaumont were indeed among the best the Signal Corps enjoyed. But like most things in war, they were temporary.[35]

Pershing's thoughtful inspection probably went beyond an attempt to forestall disrespect for female units. The switchboard unit was a critical cog in Pershing's war machine. He may have wanted to guarantee that it was well oiled. Pershing may also have wanted to ensure he could field any civilian complaints that women were not being treated properly. Scandal could damage the war effort.

Like other soldiers who took obvious pleasure in the presence of "real American girls" who reminded them of home, Pershing also appeared curious to see the unusual recruits close-up. For the general, this probably had an additional significance. When stationed on the Mexican border three years earlier, Pershing had endured an excruciating tragedy. A house fire in the officers' quarters at San Francisco's Presidio killed Pershing's young wife, Frankie, and their three daughters, the oldest age eight, the youngest only two. Pershing would wait another thirty-one years to marry again. The adoring "girls" of the Signal Corps likely stimulated his protective instincts.

Banker admitted to her diary that she was as starstruck as the old cook "ready to burst with pride" at the attention of the commander in chief. To his inner circle, Mitchell Yockelson writes, Pershing was "sensitive, warm, and caring." He certainly gave that impression to Banker and other operators.[36]

A few months later, Pershing unexpectedly sought out female troops in Tours as well. An order came one day to marshal for inspection. General Pershing walked down the line, shook each woman's hand, and emphasized that they were doing an important job. "He was tall and good looking, and had a lovely smile," one later recalled. His uniform "was a perfect fit; in fact, it looked as if he had been poured into it. I never saw another soldier who looked better dressed."[37]

At an operational level, Pershing and other officers of the Signal Corps understood that field conditions, adequate accommodations, and esprit de corps would shape operators' ability to function. To prepare for "the big push," when the United States would finally take the offensive, soldiers must have their bivouacs in order. Settling into army life was thus each woman's third major task.

Although Signal Corps operators needed the same services as other troops, the army was initially not sure how to provide them. After working long shifts, operators could hardly be expected to cook and clean like "regular" women, yet they also could not be expected to live like infantry. How might the army accommodate troops who were neither "officers" nor "men," the only types of human that military organizations normally recognized?

General Edgar Russel decided that operators should be housed together. It was not conceivable to allow such vital personnel, engaged in manifestly sensitive work, to construct their personal lives however they wanted in the manner of civilians. Yet Russel also did not want male officers prying into the women's affairs, nor did overburdened quartermasters have time to equip special barracks in France. Who could look out for a tiny group of 223 females in an army that would grow 2,000 percent in two years? True to form, the War Department ignored the question until the last moment.[38]

The admixture of public and private initiative came to the rescue again. Private organizations like the YMCA had stationed themselves throughout France to provide wholesome recreational activities for soldiers—a self-imposed task that the Christian organization had performed for the U.S. Army since the American Civil War. The Y operated "huts" that passed out sandwiches, served coffee, poured lemonade, and provided "clean" entertainment for men off duty. The Salvation Army became renowned for the hot doughnuts served by its "lassies," and the Red Cross for aid to military hospitals and French orphanages.[39]

The induction of women created a unique role for the female version of the Y—the YWCA. In March 1918, the Signal Corps sent an emergency request. A harried army officer (unnamed in YWCA records) descended on the Paris office, frantic at the imminent arrival of the first contingent. Would the YWCA be willing to set up billets, chaperone social engagements, and supervise the general physical welfare of the operators?

The anxious army officer made plain that the women were pioneers. Their treatment and conduct would be scrutinized. They "might merely be the fore-runner of an organization of women such as the W.A.A.C. with the British Army." U.S. forces were expanding so rapidly that full-fledged induction of women seemed a real possibility.[40]

The spokesman "took his responsibility very seriously," a YWCA report noted, "and was manifestly cheered when reassured that the Young Women's Christian Organization would take care of that and any succeeding units sent over."

In their memorandum of understanding, General Russel put the Christian organization on notice that the army allowed no "religious ceremonies" or "discussion of religious matters."[41]

Members of the YWCA embraced the opportunity to make an enduring mark. Proud of the assignment, the charitable organization featured Signal Corps women on dramatic stateside posters that asked Americans for funds to "Back Our Girls over There." As one volunteer articulated her perspective, "One of the things to be accomplished is perhaps, for the YWCA over here, to make such a favorable impression on all American women, that they will know that when they return to the United States it will be there to . . . continue to work with them."[42]

YWCA secretaries, as they were sometimes called, prepared themselves "to endure 'hardships of war' in shot and shell if need be," as one described her mind-set. Another wrote, "It is a great privilege to make a home for these girls . . . for their work is very nerve-wracking."[43] YWCA organizers dashed ahead of the operators to rent homes and hotels, bargain with landlords, and haggle for supplies. From March to November 1918, they fumigated, rehabilitated, furnished, decorated, and hired staff for twelve "Signal Corps houses." The largest facilities, in the Atlantic seaports, were "Army Hotels for women," which accommodated the Signal Corps plus female civilians that the Ordnance and Quartermaster departments brought over as clerks.

From Brest and LeHavre on the coast, to Chaumont and Neufchateau near the front lines, YWCA volunteers put themselves at the service of the women they called their "family" and admired as "splendidly alive and very interesting." They counted themselves successful at moments when they created peaceful, happy interludes where "we *all* forgot" the war."[44]

Chatty YWCA reports of the period reveal an effort to blend billeting regulations with the expectations of American domesticity. Fresh wallpaper, flowers on the table, and Sunday teas reassured visitors that female soldiers had not discarded their femininity along with civilian dress. The YWCA strove to make army quarters "as homelike as possible," sustaining operators' morale and the organization's middle-class notions of correct deportment for women.[45]

Accommodations varied widely. Grace Banker's unit at Chaumont lucked into a three-story stone house previously furnished for officers. It possessed three bathtubs, and the resourceful YWCA hostess conjured a piano—all great luxuries. But when the front moved several months later, the unit found itself in a wooden shanty on cots with straw pallets. In Ligny-en-Barrois, Berthe Hunt and another operator shared a bed in a garret crawling with bugs. Louise LeBreton occupied a room with two mattresses and five women. In ports and other cities well behind the lines, YWCA secured entire hotels for the operators, two to a suite. Even there, however, overcrowding meant "all too frequently two women have to share one bed."[46]

Signal Corps operators appreciated the help while resisting secretaries who treated them as young girls instead of soldiers. YWCA volunteers sometimes came across as prudish. Merle Egan's "house mother," as she dubbed the secretary, "labored under the delusion that we would be molested by an ardent Frenchman so she insisted we carry either a long slender cane or a riding whip for protection." Egan ditched the silly cane after her first outing. The Signal Corps expected the YWCA to enforce such rules as dismissing guests by 10:00 P.M. and turning out lights by 10:15 P.M. "Can't go out in the streets unless accompanied by another operator," thirty-three-year-old Berthe Hunt complained to her diary.[47] Yet a private organization had no real authority over army personnel. Without rank, persuasion was the YWCA's only tool.

The YWCA finessed the relationship better than the male version of its organization, which sometimes came across as condescending and even hostile. Ironically, females who worked for the YMCA were among the worst, perhaps because they felt their service to men made them more important. "We don't want you girls in here," one matron publicly chastised an operator who dropped by a Y hut on a business errand. "This is for men only."

Class antagonisms occasionally intruded as well. When YMCA officials used a Signal Corps house for a meeting, Grace Banker found she practically had to force her crew to talk with "the old pills" who "looked us over in a most patronizing way" as the lucky recipients of "charity for the working girl." When YMCA biddies gave Banker a bouquet of forget-me-nots, she wrote in her diary, "Forget me not? You bet I won't. No one is going to look down on my girls."[48]

When the first Signal Corps contingent arrived, apprehensive YWCA hostesses feared the army might have selected low-class "girls with blondined hair and who would be dressed accordingly." They were relieved to discover that recruits seemed to share their middle-class sensibilities—or at least appeared to do so with their regulation "blue uniforms, plain blue hats, [and] neat black shoes." Gender solidarity also mitigated class differences. Their sex made all women invisible and unreal to some men in charge. Officially—and bizarrely— as one YWCA official in Paris noted, "the War Department still takes the attitude that women are no[t] in the war, and that women are not in France and will not be in France."[49]

The matrons' respect grew from sharing the workingwomen's hardships. As a YWCA organizer reported, "For the most part I am just here trying to be an unobtrusive but watchful chaperone," especially when male officers visited the house. Another admitted, "I realized almost at once that the girls resented the idea of being 'chaperoned' and I resented being only a housekeeper. The only solution was, as nearly as the girls themselves would permit, to become one of them."[50] When operators fell sick with influenza at the height of the epidemic, YWCA hostesses nursed them through the crisis.

Although secretaries sometimes complained of having the responsibility of mothers and military officers with the authority of neither, many also felt humbled by the operators' contribution to the war. "To see them sit at the board and work against great odds and with such a noise all about them, to get an important call through, which perhaps might mean a troop movement order, makes one realize why a girl comes home tired and with nerves on edge, after eight hours of such a strain." The antidote, they found, was "fun"—and lots of it.[51]

YWCA secretaries took entertainment seriously. Laugher and gaiety were tonics. Soldiers under pressure needed to shake off stress as quickly as possible to retain effectiveness. Accumulated tensions heightened fatigue and increased the risk of mistakes. YWCA hostesses organized innumerable parties at the women's residences. "We have had three parties this month," the YWCA secretary in Brest reported, "real parties with invitations." Other army units and private organizations also invited the women to fetes, where, the *Philadelphia Inquirer* reported, "there are so many more men than girls . . . that the girls are called upon not only to 'halve,' but to 'quarter' their dances so that every man may have at least one dance or rather be part of a dance."[52]

As the operators established their place in the army, restrictions against socializing loosened, though limitations on fraternizing with enlisted men remained strong in some settings, mostly "because we might lose our status as semi-officers," Banker noted.[53] For young unmarried soldiers, the chance to interact in such an adventurous context gave their service a special savor. Military life leveled the playing field between men and women in new ways. The operators reveled in their independence.

"The house is filled with officers every night," the Brest YWCA secretary observed. When one party was over, the occupants declined assistance with putting tables and chairs back in order: "All of the girls helped, and in ten minutes we had moved the grand piano back into the salon and had brought all of the salon furniture downstairs again." The young women and their hostess "laughed until we cried almost." Officers who came back later were "astounded and disappointed not to help us the following day, but the girls thought it a great lark."[54]

Romantic encounters resulted as well. For heterosexual women in the Signal Corps, it was undoubtedly agreeable to receive so much positive male attention. Officers of all nationalities vied for dances at parties, enlisted men expressed poignant appreciation when included, and hospital patients lit up when operators visited, as many of them did when off duty. Colonel Parker Hitt wrote his wife, "I am responsible for getting the sweet young things over here to give us some real American telephone service and our gay bachelors are duly grateful for other reasons."[55]

American bachelors had some competition. Ellen Turner, a Seattle recruit, wrote her family from Le Havre, "I thoroughly agree with Marie who used to say that there is nobody in the world who can be nicer to girls than Canadians. Just at present I am rushing the Canadians very strong." But

American officers "rushed," too. Lieutenant Marmion Mills, one of Pershing's staff officers, happened by the telephone exchange more frequently than army business seemed to require. He and Adele Hoppock enjoyed a long life together after the war. Lieutenant Riser had eyes for a Massachusetts soldier. "Chaperoned Lt. R. and Miss Langelier at dinner in the village Hotel de France," Grace Banker noted discreetly in her diary a month after arriving in Chaumont.[56]

An "awfully good looking" Australian officer with elegant manners and a heroic war record attached himself to Banker, who found herself missing him whenever he went to the front. Captain Mack exhibited the mating behavior unique to a war zone, giving Banker tours of experimental gas fields and sheltering her under his arm when the shells of a trench mortar "exploded and fell all around us." The Australian took the chief operator horseback riding along a quiet section of the Marne. Both were disappointed when a YMCA director failed to seat them next to each other at a formal dinner.[57]

Captain Mack, whose first name Banker never recorded, returned to the front wearing Banker's Signal Corps pin. When he came back decorated with the American Distinguished Service Cross, he gave it to Banker to wear. "I have no right," Banker wrote, conscious of the far greater dangers he braved—and perhaps the obligation of an unmarried woman to refuse valuable gifts from a single man (which she clearly did not observe).[58]

Male admiration was familiar to women who enjoyed the unearned advantage of good looks, inherited like money and just as likely to create a sense of entitlement. For plainer women, the experience was novel. Average-looking Merle Egan took the train out of Montana committed to a man who had made no commitment in return, supremely conscious of her "unsettled love life."

Egan recognized that to "lonesome men overseas, an American woman that wouldn't have caused a second glance at home became the object of pent-up emotions." Nonetheless, her memoir after the war revealed delight when soldiers of all ages sought her out, some clearly infatuated. (Egan followed army regulations and consigned her diary to the Atlantic on the way to France.) Bright-eyed and confident, the dark-eyed Egan attracted considerable attention in her fitted uniform. Dinners at the Officers' Club, strolls in the moonlight, splurges on artichokes and lobster, and afternoons spent wandering French avenues hand in hand opened her eyes to a fuller, more exciting life than the one she had left behind. "I received my first good night kiss from Randie and I enjoyed it," she noted with satisfaction one evening.[59] Army life was rarely so pleasant.

For one soldier, it ended more abruptly than anticipated. Without revealing her name, Signal Corps operators in later years recalled the sole woman court-martialed and sent home in disgrace. Merle Egan told attorney Mark Hough that the soldier had gotten pregnant, an alarming event given the spotlight on female pioneers and the inability of women to hide conception. Egan may have been speaking of Lucienne Bigou. Captain Wessen had been cautioned about placing the eighteen-year-old within reach of unladylike "temptations." Personnel records did not reveal why the army expelled her two months after the armistice, but it may have been a consequence of the revelry that followed victory. "Bigou has been relieved from duty with the AEF and directed to proceed to US for discharge on account of inefficiency," Lieutenant General James Harbord opaquely cabled the War Department.[60]

Overall, the YWCA provided an extraordinary service to the female operators and thus the American Expeditionary

Forces. Dances, parties, and clean accommodations proved invaluable. The YWCA saved the army from its own "misman-agement" of female personnel, in the view of YWCA host-esses. One secretary working in Le Havre believed this was due "to no lack of interest on their part, but to the general con-dition of unpreparedness in the States when war was declared and to the consequent 'speeding-up' of ways and means." Like other Americans, she looked with envy upon the superior organization of the English WAACs.[61]

Signal Corps houses also provided the setting for the fourth and final task of new recruits. Like doughboys, they needed buddies. Written shortly after the war, the 1922 song "My Buddy" expressed the nostalgia that most Americans in the AEF ultimately felt towards fellow recruits who helped them through that difficult time: *"Buddies through all the gay days, Buddies when something went wrong; I wait alone through the gray days, missing your smile and song."*

Few people traveled any distance in this era, especially young women. They presented a brave face in their jaunty let-ters home, but Berthe Hunt's diary—written for her husband, if he survived—reveals the loneliness that beset everyone. Once they made the adventurous decision, many women found themselves stricken with homesickness and fear. "Some of the girls are having the blues & breaking down but that is to be expected," Berthe Hunt wrote not long after her May arrival in Neufchateau, the telephone exchange then closest to the front. Midnight air-raid alerts, German warplanes, lights-out after dark under penalty of imprisonment, and gun flashes on the horizon made war real. Death was never far away. "Oh, those thuds go right through you," Hunt admitted.[62]

"The new girls came today," she wrote when the fourth unit arrived. One of the women was "very blue" but able to "blurt things out," which helped her regain her composure. Another was determined to be a "good soldier" and never made a murmur. Both expressed dismay at the harsh living conditions. The veterans of Neufchateau recognized their distress. "We feel sorry for them & know their feelings for we have been there," Hunt wrote. Her own diary frequently reflected worry, sadness, and frustration. "A desperate attack of lonesomeness," a typical entry read in late July.[63]

Comradeship held them together. At Chaumont, the women hiked three miles each way to work, keeping one another company in the early morning and after dark, when they struggled along muddy roads. At Ligny-en-Barrois, also close to the front, they slept on the same mattresses and fought insects together. "Bed bugs with us still so we were on the hunt a good part of the night," Berthe Hunt wrote in September, when transferred there. The next day, she and Marie Lange bonded over insect powder and turpentine. Lange had a "good heart," Hunt felt.[64]

Hunt, LeBreton, Banker, Hoppock, and others spent rare leisure hours hiking, picnicking, or bicycling together in the French countryside. In one misadventure, Hunt and Marjorie McKillop from the University of Washington discovered flimsy brakes on their rental bikes. Undeterred, they jumped off before each downhill until one caught them by surprise. "Coming down, our bikes ran away from us & Miss Mc steered into a field of wheat & when I turned the bend all that could be seen was her head & shoulders & arms gesticulating—how we did laugh!" When her roommate was transferred a month later, Hunt despaired. "I hate to think of the days without her."[65]

In addition to billet mates, the Signal Corps operators enjoyed the occasional companionship of French and British operators with whom they felt a special sisterhood. "Met British telephone operators [WAACs] & were treated most courteously," Hunt wrote the afternoon she rode her bike to a town where the British Army had its regional headquarters. Another day she took tea with "French military telephone girls," who were "enlisted" just like Signal Corps operators, Hunt wrote, reflecting her ignorance of the U.S. War Department's less generous view of her status.[66]

Strangely, only American nurses proved immune to friendship. When Merge Egan was left behind in England for a week, the army placed her with a group of them. The nurses shared their rations and hard cots with Egan but were "too tired or too busy" for conversation. When Egan crossed the Channel with yet another contingent, the nurses "ignored me so completely that I felt like an outcast."[67]

Berthe Hunt got the cold shoulder as well when visiting the hospital near her post. Despite a common cause, "American nurses are snobs . . . cool as can be to us—you would never think we were American women far from home & all in the same work."[68]

Army nurses made their displeasure known to Grace Banker the first week of her arrival in Chaumont. Banker "tried to speak to them on the street one day," greeting them as fellow Americans, but they were "not very cordial." The nurses "did not seem to care about it," Banker observed with surprise. When the nurses dropped by the Signal Corps house six weeks later to repay a courtesy call Banker and her YWCA hostess had made, it was "a cold formal" occasion. The nurses stayed "four and a half minutes to be exact," Banker noted. Another time, when politeness required Banker to accept an invitation to a dance at the base hospital, she noted that "the nurses simply looked through us." Banker wrote:

It is the queerest thing. They just do not like us. Are they jealous? Of course we are younger but then we are all American girls. How petty! Our girls are for the most part well educated, some college, some private school, and all have good spirit. The wounded boys who come down from the hospital all say, "Gee how those nurses do hate you."[69]

Lacking nurses' testimony, it is not possible to know why these older professional women collectively disdained the Signal Corps operators. Grace Banker's comment implies she thought it might be class hostility. Perhaps nurses felt they were above "working girls."

Nurses may have envied the privileges granted the tiny corps of operators, who enjoyed billets similar to officers. Most of the five thousand army nurses who served in France lived in wooden barracks or, worse, cold, drafty tents attached to vulnerable field hospitals. Blackout regulations meant stumbling around after nightfall. One nurse near the front kept a small pail of water under her cot to make a cup of tea in the dark after her shift and another pail to relieve her bladder. A tent mate later recalled, "One night we were awakened by her cussing—she had mistakenly used the wrong bucket for her tea." After a reception the YWCA gave for nurses at the Brest Signal Corps house, one said wistfully upon leaving, "This is the only real home I have been in, in almost sixteen months."[70]

The nurses may also have been possessive of their status as the only females with permanent standing in the U.S. Army. They were a far larger unit, some performing more hazardous though no less essential work. Who were these new Hello Girls to snatch their glory? The War Department's failure to welcome the operators officially, as the navy had the yeomanettes, made legitimacy a limited, contested commodity.

War Department intransigence heightened status conflicts among operators, too. Unlike males with officer training, Signal Corps supervisors had little preparation, minimal guidance, and ambiguous authority. As the befuddled commanding officer aboard the SS *Carmania* observed with regard to Inez Crittenden, neither she nor women under her could produce orders verifying Crittenden's command. "Notwithstanding this rather peculiar situation," he reported, "the members of the telephone unit have, without exception, accepted the control . . . as legitimate."[71]

Yet the lack of clarity undermined authority and may have caused some women to go too far in asserting their command. Female officers needed greater tact than males, whose ranks were absolute rather than "relative." Unsurprisingly, supervisors did not all prove equally adept at managing their ill-defined responsibilities.

"Oh, I was mad this morning," the normally good-natured Berthe Hunt wrote one day after she told her chief operator about a talk she had with other women about household tasks. To the married woman's surprise, "that made the C.O. mad—told the girls I had gone over her head, was liable to court martial, she was *dictator* in every detail (home as well as office)." Hunt knew there was nothing to be done about such high-handedness, but she was furious.[72]

Both Hunt and her supervisor occupied untenable positions. The Signal Corps made chief operators responsible for troops twenty-four hours a day. Male officers had a similar responsibility for enlisted men, but they did not live with them. Black Jack Pershing and General Edgar Russel were not expected to take tea in the parlor, exchange embraces, or act as "family" in the manner expected of women.

Grace Banker understood the challenge and impossibility of being both friend and commander. She waited up nights until her charges were home ("one of my jobs"), encouraged them to write their mothers on Mother's Day, spied on an operator whom G2 had asked her to watch, accompanied Catholics to mass and Protestants to services, and enforced military regulations. In Grace Banker's private journal, she called them Tootsie, Hoppie, and other nicknames, but noted, "I call them by their first names in my diary only. Can't be too familiar." Only Berthe Hunt, who must have seemed an ancient married person, appeared as "Mrs. Hunt" even in Banker's private notes.[73]

One woman with whom Banker had repeated trouble was nineteen-year-old Louise LeBreton. Their relationship reveals the challenges women faced in a military context. Lacking indoctrination in boot camp or grooming in officers' school, some women underreacted to military discipline while others overreacted. Even the skillful Banker struggled—though her actions showed that stern discipline shielded junior operators from worse measures by male superiors.

"Louise comes from the University of California," Banker wrote in her diary. "I like her."[74] Generous spirited, LeBreton shared with everyone the jam and crackers that her mother had packed. Yet the nineteen-year-old masquerading as a twenty-two-year-old could be irrepressible. Banker sharply disciplined the Californian at least three times. Louise viewed Banker as a stern taskmaster, especially during their initial weeks in Chaumont.

On the first occasion, Louise broke the rule that chief operators must maintain "order, quiet, and discipline" in the office at all times. Without warning, a red light on Louise's switchboard showed someone on Pershing's line. Every jack on the switchboard was taken. Knowing the commander had precedence over all other callers, she yanked the plug from someone else's call. "Number, please?" she asked, determined to sound professional while squealing inside. An authoritative voice inquired, "What time is it operator?" LeBreton was instantly flummoxed. Surely the great man needed someone important, like the president of France. She stammered, "I beg your pardon, Sir?"

The voice replied, "Operator, this is General Pershing. What time is it, please?" LeBreton looked at the clock over the door and answered, "It is 9:20, Sir!" The caller thanked her and hung up. All would have been well had Louise not sang out above the hubbub of other women connecting calls, "General Pershing just asked me for the time!"

Soldiers did not yell across a military exchange like children on a playground or girls at a party. Banker reacted swiftly. "Follow me, Miss LeBreton." Outside the hearing of others, the angry chief operator confined the young woman to quarters for thirty days when not on duty for spreading a rumor about Pershing, who was not in camp. When Banker later found out that Pershing had indeed dropped in unexpectedly, she reduced the sentence to a week for disturbing the office.

LeBreton soon got herself into trouble again, "quite innocently," she thought. In a letter home, Louise employed French terms to hint at where she was stationed, a violation of military secrecy. The censor tore up the letter. Banker issued a reprimand. A few weeks later, Louise heard a new office might be opened at Langres, south of Neufchateau, and wrote her sister Raymonde in Tours to request a transfer. Perhaps they could be together.

As Louise well knew, soldiers were only allowed to write they were "somewhere in France." She had violated censorship rules again. Lieutenant Riser notified Grace Banker, "this is the second offense of this young lady. . . . You are, therefore, to inform Miss LeBreton that she has her choice of company punishment or trial by Court Martial." Two months into the war, Banker faced her first serious discipline crisis. LeBreton might be the first Hello Girl sent home. "If she decides to accept company punishment, it will not be necessary for her to stand trial," Riser wrote.

Grace Banker must have sternly counseled the Californian. She reported back to Lieutenant Riser, "Miss LeBreton prefers to take company punishment." Once again Banker combined severity with clemency, confining LeBreton to quarters for thirty days but suspending most of the sentence "pending future good behavior." Banker wanted to train the younger woman, not break her spirit. LeBreton was an adept, valuable operator, eventually promoted to supervisor and transferred to the edge of battle during the first American offensive. Banker had less patience with women unwilling to learn, such as a Seattle co-ed whom she considered "peevish and mean and can't obey orders or take corrections."[75]

LeBreton's next brush with a court-martial resulted from adherence to the rules Grace Banker ground into her. On this occasion, protocol protected LeBreton, as it was supposed to. The location of armies, corps, artillery, and divisions was strictly secret. When a subscriber asked for a town without using the correct code, one operator later recalled, "we were compelled to reply that we had never heard of such a place!"[76]

One night when the war lit up the switchboards, Louise LeBreton fielded a request from an irascible officer who demanded to be connected with Pershing's advance headquarters. Louise asked for the code, to which he barked, "You know very well where it's located, operator." When she replied, "We are not given that information, Sir," and asked again for the code, he demanded her name. When she answered, "I am not allowed to give you my name, Sir," and told him she was operator 22, he insisted on speaking with a man. The next day, LeBreton was summoned to a possible court-martial for insubordination. When she explained what had happened, the chief signal officer for Neufchateau exonerated her: "You were absolutely right and you have nothing to fear."[77]

"Military discipline must be maintained," LeBreton later said, harboring no grudge toward Grace Banker, whom she admired as "an excellent Chief Operator, fun loving but firm in her orders which we obeyed to the letter." Other operators felt the same, as Banker discovered when she left most of her unit in Paris to join Pershing. "I was touched when the girls begged to kiss me goodbye," she wrote. Banker learned three months later that some had pleaded with senior officers for a visit from Miss Banker at their outpost in Neufchateau. When Lieutenant Riser obliged by taking Banker there, "Miss Hunter hugged me and lifted me right off the floor!" To the pioneer chief operator, it felt as though her girls would always "belong" to her.[78]

Banker's immediate supervisors and subordinates were not the only ones to observe the New Jersey native's knack with running an efficient operation. General John Pershing must have done so as well, or he would never have taken her with him when U.S. troops finally assembled to move en masse. Pershing fired fourteen hundred officers during the war, from second lieutenants to major generals.[79] Only nervy, competent men and women were wanted for key positions.

In September 1918, the AEF was ready at last to take the offensive against Germany. Black Jack would first have to get approval from the same French and British commanders he had frustrated for more than a year. Foreign leaders continued to doubt what contribution an American army would make. Fortunately, Pershing could point to the bravery, commitment, and skill his soldiers had demonstrated shortly before, during Germany's final great offensive.

Back our girls over there
Y.W.C.A.
United War Work Campaign

In urgent need of experts, General John Pershing finally ordered the War Department to ship uniformed women "over there." Every operator swore to serve for the duration. The Army recruited the YWCA to arrange their billets. ("Back Our Girls Over There," Poster Collection, US 477, Hoover Institution Archives, 1918)

Forbidden to salute and paid less than men, thousands of British WAACs served. When bombs killed nine in France, Queen Mary became their sponsor. (Imperial War Museums, Art.IWM PST 13167, 1918)

Braving winter's cold, militant suffragettes picketed the White House in 1917 to shame Woodrow Wilson before the world. (Courtesy DC Public Library, Washingtoniana Division)

Dora Lewis, age 56, upon release from jail in 1918. A hunger strike leader, she was arrested four times and sentenced to eighty-eight days for picketing the White House. (Courtesy DC Public Library, Washingtoniana Division)

Embracing the persona of the "New Woman," Hope Kervin flooded
the army with letters. She wanted her own "little whack at a
German"—and lied about her age to get in. (Application photo, 1917,
NPRC, Margaret H. [Hope] Kervin)

In Inez Crittenden's application photo, a candlestick phone sits behind her. The peppery Californian led the second unit to France, where she died "in the service of her country." (Application photo, 1917, NPRC, Inez Crittenden)

Joan of Arc escorts the third unit in a San Francisco war bond parade. Adele "Hoppie" Hoppock stands far right, wearing a white hat. Next to her, barely seen, is Berthe Hunt. (Courtesy AT&T Archives and History Center)

Women drilled atop AT&T's Manhattan headquarters. *L–R:* Suzanne Prevot, Charlotte Gyss, Esther "Tootsie" Fresnel, Fernande Van Balkom, Grace Banker, and Frances Paine. (Courtesy Robert, Grace, and Carolyn Timbie)

After a harrowing aerial bombardment, the first unit posed for their picture, eager to begin duty. Chief Operator Grace Banker sits in the center of the first row. (NARA, RG111, Entry 45, Box 400, 111-SC-8445)

"Bell Battalions" lifted from AT&T's domestic workforce installed telephones in trenches near the Somme battlefields, March 1918. (NARA, 111-SC-8379)

Signal Corps linemen repair broken telephone wires under poison gas attack. (NARA, 111-SC-159447)

Sitting near a poster of Joan of Arc, Raymonde Breton (R) visits her sister Louise in the Signal Corps barracks at Neufchateau, later bombed by German planes. (NARA, 111-SC-50699)

L–R: Berthe Hunt, Tootsie Fresnel, and Grace Banker run Pershing's switchboard at First Army headquarters. They keep helmets and gas masks close. (NARA, 111-SC-21981)

In a rare break, officers and operators celebrate four birthdays.
Far right, in profile, Colonel Parker Hitt. *L–R:* operators
Berthe Hunt, Suzanne Prevot, Adele Hoppock, Tootsie Fresnel,
and Helen Hill. Grace Banker sits far left, with only her sleeve
visible. (Parker Hitt Collection, U.S. Army Intelligence
Center of Excellence, Fort Huachuca)

Fire engulfs the wooden Signal Corps barracks when a German
POW overturns a stove. Operators keep lines open to the front
until the last second while soldiers rescue personal belongings.
(NARA, 111-SC-28408)

In Paris to handle communications for the peace talks, Grace Banker and Merle Egan pose atop a captured German cannon in front of the Hotel de Crillon on the Place de La Concorde. Banker dubbed herself a conservative, while Egan joked that she'd become a Bolshevik. (Courtesy Timbie family)

General John "Black Jack" Pershing inspects operators
serving in occupied Germany. Women remained on duty until
discharged. The last were relieved in 1920. (NARA, 111-SC-156403)

Three service stripes on her sleeve, Grace Banker wears the
Distinguished Service Medal, awarded to only eighteen Signal Corps
officers of the U.S. Army, including her.
(Courtesy Robert, Grace, and Carolyn Timbie)

BELL TELEPHONE NEWS

VICTORY
hath crowned the
valor of thy sons.
O America, who
now lay by thy
sword unsullied
and turn again
to paths of peace
beneath thy glor-
ious citizenship.

VOL. 9. NO. 1   AUGUST, 1919

AT&T honored operators and doughboys in 1919. The
campaign to ratify the Nineteenth Amendment and grant full
citizenship to women was still underway. (Courtesy of
AT&T Archives and History Center)

Merle Egan Anderson (age ninety-one) and Mark Hough (thirty-four) fought the army and won. Hough said there are some people in life "you bond with." (Museum of History & Industry, *Seattle Post-Intelligencer* Collection, 2000.107.008.6.01)

8

BETTER LATE THAN NEVER
ON THE MARNE

To most operators, the war felt eerily distant.

Outside Paris, with its draped windows and ruined buildings, and beyond what the army called the Advanced Section, where operators worked right behind the front, large swaths of the ancient countryside felt almost too normal given volunteers' passionate desire to help win the war.

Many men felt disconnected as well. Of the two million soldiers who served in France, roughly 600,000 worked behind the lines in transport, communications, construction, engineering, road building, and other noncombat jobs.[1] General John Pershing realized early in 1917 that this required tact. Most volunteers wanted to prove their bravery. (Draftees may have felt otherwise.) Rather than speaking of soldiers at "the rear," Pershing designated troops assigned to jobs other than combat as "Services of Supply," or SOS, evoking the signal for "help" in Morse code. Even so, Pershing wrote, "the idea prevailed in the minds of its personnel . . . that they were not exactly doing the work of soldiers, and hence, their efforts lacked enthusiasm."[2] He took every opportunity to buck them up.

SOS made its headquarters at Tours, southwest of Paris on the road to the port of Saint-Nazaire. Switchboard operating there required adjusting one's expectations. "In Tours I felt further from the war than I did in Helena," Merle Egan admitted. In Montana, she had avidly purchased war bonds. In France, she bought souvenirs. Back home, she had read the newspaper and skimped on flour to conserve food for the troops. In the SOS, army censors filtered war news and rations were plentiful. Helena had a "feverish war atmosphere." Sunny Tours was serene. The air-raid alerts, sooty skies, wounded soldiers, shell-shocked civilians, POW camps, and artillery concussions of the front were two hundred miles away.[3]

Even work followed an eight-to-five schedule. Female operators took daytime shifts when busy switchboards required fast hands. Men staffed slower overnight shifts. "Our only contact with the war was when we handled long distance calls from the front," Egan recalled. "Then, some harried officer, trying to reach Service of Supplies for some needed equipment, would be so near the front we could hear the roar of guns in the distance."[4]

Many women wished they could serve nearer the action. "I'm tired of being a feather-bed soldier," Ellen Turner wrote her family from the port of Le Havre. She hoped to be "lucky enough" to be ordered to the front. When Adele Hoppock was sent to Neufchateau, she wrote from "somewhere in France" that other girls envied her "because it is near the front." Hoppock downplayed the danger and told her family, "It is the place above all others that I would choose. I feel very fortunate."[5]

Yet combat forces achieved nothing without the Services of Supply. Logistical troops constituted the foundation that Pershing had spent a year building. Army discipline meant accepting that one must go wherever sent, to do whatever

commanded. Like other soldiers, Signal Corps women understood this. But they considered themselves privileged if they
were among those who followed Grace Banker and Black
Jack Pershing.

Other women looked upon them as heroes, much as male
soldiers who never crossed the Atlantic later viewed recruits
like Alvin C. York, a recipient of the U.S. Medal of Honor and
French Legion of Honor who became the face of the doughboy
after the war, when the *Saturday Evening Post* lionized his
exploits.

Alvin York exemplified what America believed about itself.
The thirty-year-old backwoodsman from Tennessee did not
cotton to war and had done everything to avoid it. A self-
effacing Christian fundamentalist, Alvin York twice applied
for conscientious objection. But when turned down, he duti-
fully enlisted. On an October mission in the Argonne Forest,
York's unit found itself surrounded. Every superior officer was
killed or wounded. Leadership fell to the sharp-shooting
Tennessean. Crawling up a mountainside, Corporal York
single-handedly silenced the machine gunners decimating his
unit. By the end of the day, he had killed more than twenty
Germans and taken 132 prisoners. Every male volunteer
wanted to be an Alvin York—and even if he couldn't, the Tennessean's fame made each feel part of a heroic campaign.[6]

Similarly, most female volunteers wanted to serve as a
member of an advance unit. And if they couldn't, those who
did proved the bravery and good citizenship of which all females were capable. It was almost as good as being there, to
know "our girls" were giving their utmost to the struggle to
which other operators contributed in humbler ways. Chief
Operator Nellie Snow, in charge of a quieter exchange than
Grace Banker's, commented that while she had hoped to

serve within shell range, even those who looked on from afar were "the happiest women in the world, for we were allowed to come to France to do our part in the winning of the war."[7]

Grace Banker's own chance to approach the front came after America's opening engagement in May 1918.

———

"First very important message through & I had to relay it," Berthe Hunt recorded cryptically at Neufchateau on May 27, the day the army's First Infantry Division moved into position on the eve of the Battle of Cantigny to stop the enemy from advancing deeper into France. The untested Signal Corps women and the U.S. Army would both have to prove themselves to win a spot on the dangerous, complex battlefield.[8]

The Allies remained mistrustful of Pershing's larger strategy. They needed immediate help. Marshal Ferdinand Foch, appointed commander in chief of Allied operations at the nadir of Germany's spring assault, prevailed at least temporarily. Black Jack Pershing agreed to siphon off several of his strongest divisions and place them under French and British leadership to halt the German offensive in late May. The American units would remain intact, however. As Pershing later recalled, "The Allies were skeptical as to the ability of our divisions . . . to conduct an offensive."[9] Pershing wanted to prove his men's merit as a cohesive force. The same murderous spring day that the Americans moved into position, the German Army took fifty thousand French prisoners and advanced thirty miles over lands so battle scarred that they mimicked a moonscape.

Once again, long-range explosives rained on Paris. German forces imperiled the French capital. In early June, Signal Corps operators were told to prepare for immediate evacuation. The women protested that they would stay as long as the men did. "The telephone business, we felt, had become well-nigh unmanageable on account of the [German] drive going on, and it seemed to us that to put inexperienced boys in our places might prove disastrous," one Hello Girl later recalled. When shell fragments shattered a window in the Paris telephone exchange, the women refused to abandon their switchboards for the bomb shelter. "We will stay until the last man leaves," they insisted. Assuming that the operators were regular army, an admiring journalist observed, "This is the fiber of the enlisted sisters of our fighting men."[10]

Not privy to the details of grand strategy, Grace Banker nonetheless suspected the Allies were not doing well. The bright blue skies over Chaumont gave her "that lazy feeling that every thing ought to be all right," yet she sensed something was terribly wrong. She knew better than to ask. "I made that mistake only once with Lt. Riser," she admitted, who told her, "I was in the Army and there one asks nothing but keeps one's eyes open." Banker despaired at Germany's implacability and Russia's indifference to the plight of its former allies. She longed to be closer to the front, but she noted, "Every one says it is no place for a woman."[11]

While the French and British armies countered bigger attacks elsewhere, American divisions fought the first of three small offensives to halt the German march toward Paris. Each engagement bolstered Pershing's claim that America could field an effective force. The first took place at Cantigny, an evacuated farming village on a ridge the Germans had taken and from which they could rain artillery fire down on the surrounding area. An American division wedged itself into the line held by the French, with orders to take the hill town.

At sunrise on May 28, 1918, guns loaned to U.S. forces opened a rolling barrage over the heads of crouching troops. Smoke roiled up from Cantigny as U.S. infantrymen started forward in the dawn, each carrying his rifle, 220 rounds of ammunition, two hand grenades, two canteens of water, and enough hard tack and corned beef for two days. The Americans looked as if they "were at inspection," according to one observer, their lines as straight as "a long picket fence," according to another. French tanks provided cover as German machine guns began firing. In less than an hour, the Americans took the town, the ridge, and nearly three hundred prisoners. The process seemed easy until German reinforcements counterattacked.[12]

"Big drive on they say—orders are to rush messages through," Berthe Hunt wrote that day, though she had no idea where the fight was taking place.[13] Several hours later, as she watched new recruits clap, stamp their feet, and whistle at a YMCA concert, Hunt ached with the knowledge that many would soon be under fire. Meanwhile, at Cantigny, American soldiers hunkered down in bombed buildings and laid barbed wire in front of their new positions. German artillery immediately began shelling them, further demolishing the village, while enemy infantry mounted periodic counterassaults over the course of three days.

Telephones allowed the men pinned down in the village to communicate with officers and gun units behind them. One American artillery expert found himself in a stone cellar with barrels of cider, hiding among dirty infantry officers, when a field telephone suddenly rang. Surprised to learn that the call was for him, he relayed his best estimates about the angle and location of the German guns, making one among thousands of contributions to the campaign.[14]

The Germans finally fell back on May 30, leaving more than a thousand American casualties but handing Pershing his first victory. He later wrote that the successful assault at Cantigny was a source of encouragement to the whole AEF— "that the troops of this division, in their first battle, displayed the fortitude and courage of veterans." The Battle of Cantigny helped establish the reputation of the young career officer who planned the operation, George C. Marshall.[15]

The same day, Pershing ordered the army's Third Infantry Division to Château-Thierry, the next test of the green AEF. The town of Château-Thierry abridged the Marne River, a natural impediment to German advancement toward Paris, roughly fifty miles away. Buses took the Americans to their rendezvous point, where, in support of French forces, they dug in around the modest manufacturing town to keep the enemy from crossing the river.

On June 1, the oncoming Germans entered Château-Thierry. French explosives experts wired and demolished the main bridge over the river while American troops provided covering fire. When American machine gunners defending a railroad bridge came under attack, they telephoned the French artillery—again making use of Alexander Graham Bell's critical technology. The artillery responded within two minutes, silencing the guns trained on the U.S. position. For the next three days, American troops kept the Germans from crossing at Château-Thierry until the enemy called off its assault. From early June through the middle of July, U.S. troops patrolled the riverbank and fought off German attempts to cross, earning the Third Infantry Division the honorary title "Rock of the Marne."[16]

"Much activity at the front & French are terribly excited—continuous raids on Paris," Berthe Hunt wrote on June 3 from her vantage point in Neufchateau, a safe one hundred miles away. Neufchateau was only twenty miles from other German positions, however. The army began drilling Signal Corps women in the use of gas masks. Instructors at Neufchateau showed Berthe Hunt, Adele Hoppock, and others how to attach the stiff, bug-eyed contraptions, then sent them into a chamber where gas canisters had been deployed. "The mask isn't any too comfortable," Hunt wrote, "with a thing in the mouth between the teeth & around the gums, clothes pin on the nose." The women were told to take the masks off briefly to understand why they were necessary. After a few whiffs, their eyes started streaming "just as though we were crying."[17]

Two days after U.S. troops helped bring the German assault to a standstill at Château-Thierry, two regiments of Marines with the U.S. 2nd Division were tasked with the army's last and bloodiest objective that month: retaking a German position at Belleau Wood, seven miles away. Berthe Hunt again noticed the uptick in telephone traffic—as well as the lines that went dead. "They all say this is the crucial period of the war," she wrote a day into the American assault on Belleau Wood.[18]

French commanders told U.S. General James Harbord that he "must hold the line at all hazards," but also suggested his men dig another trench in the rear "just in case." Determined to show that Americans would not yield, Harbord replied, "We will dig no trenches to fall back to. The Marines will hold where they stand." Harbord echoed Pershing, who had said repeatedly that the Americans would fight openly rather than retreat to trenches. It was a strategy the other warring parties had tried, of course, with often disastrous results.[19]

Under the command of the French Army, a Marine brigade attacked the small woods on June 6. After marching all night to their jumping-off point, they advanced across a green wheat field that provided no protection against German machine guns hidden on the forested ridge. For the first time, the Marines experienced the meat grinder. A thousand men fell the first day alone.

One survivor recalled, "I noted the peculiar whispering sound again and again," as bullets dropped soldiers to his right and left. Some Marines made it to the forest, where they took cover. Others hopped into trenches to avoid the shrapnel that beheaded one unfortunate man. Shocked at seeing their first deaths, the Marines quickly became numb to the sight until, as one said, "I did not notice it anymore." A Texan observed, "I was now beginning to realize just what war meant."[20]

For the next twenty days, Marine riflemen fought uphill, from tree to tree. They could not rescue the wounded on account of the risk of losing more. Water was too scarce for anything but drinking. "I have not shaved for a month," one Marine said. The meat they had for rations was "a bit racy and [we] had our troubles accordingly," he noted.[21]

They fought on. The assault on Belleau Wood finally cleared the area of German troops on June 26, but at an extraordinary cost. Of the eight thousand U.S. Marines who jumped into the breach, sixteen hundred died and more than twenty-five hundred were seriously wounded. The French government renamed Bois de Belleau the Bois de la Brigade de Marine. The Germans were so impressed at the Marines' ferocity that they dubbed them "devil dogs." The battle stopped the enemy advance just thirty-five miles shy of Paris.[22]

Toward the end of July 1918, U.S. troops finally helped push enemy soldiers from the Marne region altogether. "We couldn't move fast enough & didn't have cords enough—it just sizzled," Hunt wrote during late July. She and others passed tidings of American successes "from one operator to another" while fielding requests from wounded men to be evacuated and connecting operational calls that conveyed military orders.[23]

The women's feeling of belonging intensified, and their officers' respect soared. When Pershing's car passed Grace Banker and her colleagues on the road one July morning, "without saying a word to each other we just turned and saluted the General," she wrote. Pershing promptly returned the salute. His aide later told Banker, "The general was tickled to death and said 'those girls are regular soldiers.'"[24]

French and British criticism of Pershing and the AEF faded. German commanders reassessed the American threat as July closed. The kaiser's chief of staff admitted that U.S. soldiers "may not look so good, but hell how they can fight."[25]

On August 4, French President Raymond Poincaré traveled to Chaumont to thank General Pershing for helping repulse the invasion. Grace Banker and other operators stood at attention, "as straight as the soldiers all around us." Poincaré and Pershing paced to a spot near where the women were mustered on the parade grounds. Accompanied by a flourish of trumpets, the five-foot-four-inch president draped the wide red ribbon and Grand Cross of the Legion of Honor over the American's head. "Then in real French fashion," Banker noted, "*le President* kissed Pershing on both cheeks." Flushing, the six-foot-tall American stooped to receive the unfamiliar, cross-cultural compliment.[26]

Shortly thereafter, Marshal Ferdinand Foch approved Pershing's request to retake territory in Lorraine that the Germans had held since 1914. "It was a moment of glory for

Pershing," historian Mitchell Yockelson writes.[27] Black Jack dubbed himself commander of the special assault force, called the American First Army, and began eyeing an attack on St. Mihiel—a "salient," or bulge, where German lines poked into territory under French control. This irregularity allowed the enemy to overlook the route to the French-held fortress of Verdun.

Pershing planned to establish a First Army headquarters close to St. Mihiel. Colonel Parker Hitt, Pershing's personal signal officer, immediately wrote General Russel. "In order to obtain maximum efficiency," he told Russel, "I desire to use women operators to be taken from those especially qualified by their familiarity with front line work and code station work." This meant bilingual women from Chaumont and Neufchateau, all "in good physical condition so as to stand the strain of their work and the possible hardships of living in the Army area."[28]

On August 24, Pershing met with Marshal Foch to plan the attack. They agreed to a date: September 12 at the latest. Still stationed in Chaumont, Grace Banker wrote in her diary that same day, "Something's up! Colonel Hitt has been sounding me out. Maybe there will be a chance to go farther Front." The following morning, yet another officer called Banker to ask her what she knew about the plan. She replied she hadn't any idea what he was talking about. "I have learned when to know nothing in the Army," she scribbled in her diary. Apparently passing yet another test, Banker was told to pack. They would leave that morning.[29]

Berthe Hunt heard rumors, too. "Oh something is up & this part is getting terribly important," she wrote when a superior intimated that women might advance "up toward the front." For both Hunt and Banker, the possibility took their minds off personal sorrows. Berthe missed Rube desperately. Her husband had been promoted to a "Suicide Fleet," which the *Saturday Evening Post* described as "the roughest job that mariners could face," as they were confronted with "desperate ordeals" in flimsy, repurposed civilian yachts guarding the French coastline. Just the week before, Grace Banker had received a telegram about her father's unexpected death. Grief silenced her diary for two days. After this, she worried more about her brother, whose letters had ceased. Was Eugene still alive, or was he buried under one of the raw crosses lining the roads in all directions?[30]

The opportunity to advance revived both Banker and Hunt. Accompanied by her two best operators from Chaumont, the chief operator was told her destination only after she was in a car on the way to Neufchateau to pick up three more women, including Berthe Hunt. Adele Hoppock almost got her wish to go, but learned at the last moment that Neufchateau could not spare her. Abandoning their trunks and other personal possessions, since there was barely room in the car for all six of them, the women proceeded to Ligny-en-Barrois, twenty-five miles from St. Mihiel, passing thousands of troops on their way to the same rendezvous.

Tired-looking doughboys brightened as the women drove through small towns stuffed with soldiers. Men waved their helmets and tin mess kits as they recognized the Signal Corps operators and gave them "an ovation of cheers everywhere," Berthe Hunt noted. The doughboys inspired Grace Banker as she and her operators moved in the opposite direction of fleeing civilians: "So we were hailed in town after town. We were the symbol of home and the U.S.A."[31]

Hunt had swung from low spirits to high spirits. "We have been kidnapped, as it were, by the 1st Army," the Californian wrote gaily. Grace Banker also used the word "kidnapped," which suggests that field officers were making decisions on the fly. Rear echelon officers did relish using Signal Corps women near the front, but General Pershing was taking them anyway. If women performed well, their role might grow.[32]

Pressure built on the War Department to induct women officially. Senior officers believed the war would last into 1919. In late August, as the need for able-bodied men surged, Major General C. C. Williams of the Ordnance Department requested fifty women stenographers to meet the "urgent need in France." Official enlistment would attract the most dedicated, "intelligent" women, Williams believed, and strengthen military authority over them.

General James Harbord, who assumed command of the Services of Supply after leading the Marines in the Battle of Belleau Wood, proposed in August 1918 that Washington immediately "organize a special section of the Army Service Corps to be known as the Women's Overseas Corps" and send five thousand uniformed females. Harbord happened to be the only major general who had risen through the ranks. A Kansas boy who scrabbled up from private, he was interested in performance, not appearances. Yet another officer in the War Department, a Colonel Ira Reeves, sent a blistering memorandum detailing the need for trained, militarized women—just like those the navy and marines possessed—to ensure the *speedy winning of the war*," regardless of how much it disturbed "the smug complacency" of Washington bureaucrats.[33]

The War Department dithered. It sanctimoniously cabled General Harbord, "The enlistment of women so as to subject them to punishment for desertion and minor offenses in accordance with Army regulations and military law would not meet with favorable public opinion." Of course, the press had described the Hello Girls in just these terms, to wide applause. The War Department instead approved the "employment" of five thousand women civilians, knowing it had broad authority over anyone in a war zone anyway. Female employees could be subjected to military discipline without military status. The War Department proposed to uniform five thousand new employees "the same as the telephone operators Signal Corps" and bill them for the cost.[34]

The executive secretary of the YWCA in France, Henrietta Roeflofs, upon whose unpaid shoulders the army had deposited the burden of billeting, jumped into the controversy. She urged the War Department to commission female officers, place them in charge of units organized like British WAACs, and expedite "enlistment of these women for their efficient control and to uphold their morale." Women should be housed together in regular barracks, not individual billets. Given wartime rigors, they needed physical training, too.[35]

In October 1918, the director of the War Plans Division of the U.S. Army notified Henrietta Roelofs that no action would be taken. Using women was one thing; acknowledging them was quite another. "The question of enlisting women for service in France has received earnest consideration but the War Department is not as yet convinced of the desirability or feasibility of making this most radical departure in the conduct of our military affairs."[36]

Field officers who relied on women's unstinting service had no idea what was happening, nor did Signal Corps operators. For both, "this most radical departure" was already underway. As operator Louise Barbour wrote her mother from France, there was nothing that made "you feel that you are a real part of the Army" like knowing that "Pershing can't talk to Col. House [Wilson's adviser] or Lloyd George unless you make the connection."[37]

French troops had tried numerous times since 1914 to uproot German forces ensconced at St. Mihiel. In a doomed 1915 offensive, France suffered 125,000 dead and wounded. To Pershing, this made the challenge only more stirring. Capturing St. Mihiel would prove the competence of the AEF once and for all and lift the morale of the Allies.

General Pershing arrived in Ligny-en-Barrois around the same time as the telephone operators. Troops were descending from all directions, on foot and in trucks. The small town filled rapidly. Pershing was quartered in a private railway car at the edge of the forest. All of America seemed to be moving into battle position. Soldiers camped in open fields spreading outward from Ligny.

"The boys no longer look like our clean happy boys," Berthe Hunt observed, "they are so brown, hard, & careless looking." During short breaks at lunchtime, operators climbed local hills to watch cannons firing on the German fortifications near St. Mihiel, five miles away. They passed binoculars to watch. The army placed sandbags around the telephone exchange to reinforce its walls against explosive shells. Operators took their rations from the same barrels as men: hard loaves of bread, hunks of meat, prunes, and coffee. Flies and yellow jackets got stuck in the jam provided for the bread.[38]

The Signal Corps assigned them to operational boards under the section of the army known as G3. From this point forward, calls would relate primarily to battlefield communications. "A lot depends on us but I am sure of my girls," Banker wrote. Not all officers shared her confidence, though. Two days after the Signal Corps women arrived, Berthe Hunt heard rumors that Brigadier General Edgar Russel, stationed out of Tours, was still "peeved about our being here." Although Russel had given Colonel Parker Hitt the go-ahead, he may have regretted it.[39]

General Pershing must have approved, perhaps persuaded by Hitt, who had convinced Russel to recruit the women a year earlier. As chief signal officer at First Army headquarters, Parker Hitt maintained warm, collegial relations with Grace Banker and other operators, particularly Suzanne Prevot—a tall, venturesome young woman from New York, who was Banker's best friend. Colonel Hitt affectionately dubbed Prevot the "wild cat" of the operator's unit. In her diary, Berthe Hunt noted that Hitt "had permission to take 6 along for the 1st Army & that is why we are here & probably will be kept with 1st Army as long as one of the girls & the Col. are friends."[40]

The controversy over their presence did not die down immediately. Two days later, as preparations for the drive speeded up, Hunt had Pershing himself on the line. Perhaps she overheard him say something to another caller, since she noted in her diary, "There seems to be a great deal of opposition to our being here so we feel unsettled."[41]

It is not clear from Hunt's brief notation whether she was referring to the army's reluctance to use women or to a similar problem Pershing was having with his French ally. Ferdinand Foch wanted to cancel the American operation. In a tense meeting in Pershing's railway car at Ligny-en-Barrois the preceding day, Foch had told Pershing that the British were making good headway to the north. There was no longer any need to attack St. Mihiel. Doing so would delay a combined offensive. The French commander of the Allied armies recommended, indeed insisted on, folding American troops into British and French positions.

Pershing could not believe his ears. At the eleventh hour, he must again fight his allies over the principle of an independent U.S. Army. Pershing wanted his own sector and his own army. If his men blended into another nation's force, any victory they achieved would not be theirs. "Despite the contribution of our splendid units whatever success might be attained would be counted as the achievement of the French armies and our participation regarded as entirely secondary," Pershing later explained.[42]

At first glance, Pershing's concern seems egotistical. A selfless contribution with no thought of recognition appears a far worthier stand. Yet Pershing operated in a world where the United States was playing a major international role for the first time. Military prowess guaranteed influence in 1918. Without it, America's vote would not count at war's end. Like women who wished to act as citizens, Pershing and Wilson must demonstrate that Americans were responsible, full-fledged members of the international system, willing and able to defend it by force of arms. The French commander did not appear willing to let them.

The United States could not, nor did it wish to, simply impose its way. War required intense cooperation, whether between American and French telephone operators unused to one another's cultural protocols or between top officers. Pershing would have to change Foch's mind. If he couldn't, the only recourse was American withdrawal—an outcome that was anathema to everyone but the enemy.

Pershing countered tensely, "Well, Marshal, this is a very sudden change. We are going forward as already recommended to you and approved by you, and I cannot understand why you want these changes." The U.S. general pushed the French officer to explain himself, and vice versa. The debate dragged into a second angry hour. Foch finally asked in exasperation, "Do you wish to take part in the battle?" Pershing replied stiffly, "Most assuredly, but as an American Army and in no other way."[43] He demanded that Foch fulfill his promise and allow the Americans to proceed at St. Mihiel.

Foch left without answering.

The next day, Pershing sought out his counterpart in the French Army, General Henri Pétain, with whom he enjoyed a more congenial relationship. Pétain helped craft a counter-proposal that Foch accepted on Monday, September 2. The U.S. First Army would still attack St. Mihiel—in hardly more than a week—then wheel sixty miles north to partici-pate in the Meuse-Argonne Offensive ten days later. To Per-shing's relief, Foch accepted the plan.

"Our commitments now represented a giant task," the American general admitted in his memoirs. Agreeing to a second major battle in such a short time meant moving 600,000 men and 2,700 large guns in only a few days. More than half would come directly from the battlefield at St. Mihiel—presuming triumph there—traveling over three narrow roads under cover of darkness. The logistical chal-lenge was staggering, but Pershing had faith in his staff. He vouched to Foch that they could pull it off.[44]

The workload for the six Signal Corps operators exploded. Preparations for a drive were enormous. Engineers tested and retested the lines of the commander in chief. A new switch-board appeared, with plugs painted white and a large letter "A" over jacks to the Artillery unit. Berthe Hunt was assigned to work it. She confessed, "It makes me quite nervous to work there for I feel the responsibility of it."[45]

For the first time, the women practiced with pistols. On a ridge outside the town, they loaded, aimed, and fired. The six operators slung steel trench helmets and flexible gas masks on the backs of their chairs. German bombers flew over, and American antiaircraft guns shooed them away. News arrived that Neufchateau, where Hoppock and LeBreton were stationed, had been bombed. A corner was taken off the Signal Corps barracks, but no one was hurt. Troops on the way to their jumping off points poured past the telephone exchange at Ligny-en-Barrois for hours on end.

On breaks, Signal Corps operators stood by the road to cheer on the men, whose grimy faces registered incredulity, then joy, when they realized that the uniformed women were fellow Americans. "We yelled ourselves hoarse," Berthe Hunt wrote, crying "Hello Boys," "Three Cheers," and "Yanks!" The switchboards kept the women busy, but everyone was keyed up, waiting for zero hour. Rain fell nearly every day. Eager, anxious, and envious, Adele Hoppock managed to catch a ride from Neufchateau for an afternoon.[46]

Finally, on the night of the attack, Colonel Parker Hitt sent Banker home near midnight and told her to come back at 3:00 A.M. Grace woke to the sound of cannons around 1:00 A.M., then again to her alarm at 2:15 A.M. She roused Berthe Hunt, Suzanne Prevot, and Tootsie Fresnel for a quick candlelight breakfast. The others awoke and begged to go, too, "but I needed them for later," Banker recorded in her diary.[47]

Men had previously taken the overnight shifts, but Colonel Hitt now put the women on twenty-four-hour duty. The six operators must rotate. Colonel Hitt reported that he took males off night shifts because they "would not be able to handle the rush of business." The women's resilience and skill had met and exceeded everyone's expectations.[48]

The first doughboys went over the top at 5:00 A.M. on September 12, 1918. Artillery had broken up some of the barbed-wire belts, but mud became yet another impediment as soldiers struggled out of shallow, rain-soaked trenches to cross a quagmire of old shell craters. The G3 switchboard flashed with orders and commands, absorbing the women's attention. Hitt returned after two hours with the news that the Americans were going strong. The women kept up their breakneck pace. "Worked hard, terribly hard," Banker wrote in her diary. She slept only two hours in the next two days. By the third morning, one of her eyes was so badly infected that a doctor had to lance it twice. Banker lost track of the days of the week.[49]

The offensive caught the German Army by surprise at a fortuitous moment. For unrelated reasons, German troops had just begun an orderly retreat when the attack started. Many of their front-line trenches were empty, and U.S. troops quickly advanced to the Germans' second line of defense. French units under American direction supported them, in a reversal of their roles at Cantigny. Tanks commanded by George S. Patton—a brash colonel who would play an even larger role in the next war—aided the assault. In some places, the German's planned retreat became a rout.

Victory nonetheless came at a price. Enemy machine gunners fired down from the woods, killing many before the three-day battle was over. More than forty-five hundred American doughboys were slaughtered in combat, and twenty-five hundred seriously wounded. The survivors took fifteen thousand German prisoners and liberated two hundred square miles of French territory, which enemy troops had occupied since the start of the war. "First Army's baptism of fire was a smashing success," writes Mitchell Yockelson.[50]

Civilians who had been trapped in the salient for four years emerged from the shell-shocked town of St. Mihiel to kiss and hug their liberators. Church bells pealed. The U.S. Army delivered the first French-language newspaper anyone had seen since 1914. Colonel Hitt drove Suzanne Prevot and Helen Hill, an operator from Connecticut, into the battered village the next day, their helmets and gas masks at the ready. Prevot and Hill reported back to Grace Banker that the town square fluttered with the red, white, and blue of the French flag on "every house, every person, girls' hair ribbons, girls' ties, etc." There were no males over the age of eighteen. Overcome with joy, a man separated from his wife and children for four years ran across an American pontoon bridge to find his family.[51]

Banker's work continued. "Never spent more time at the office and never enjoyed anything more," she wrote, jubilant at the army's triumph.[52] "My girls work like beavers." Banker's Australian suitor meanwhile returned from the line of fire. She had not seen him since Chaumont, but he had not forgotten her. Captain Mack "must have been a Sherlock Holmes to find us," Banker wrote, likely wondering at the extent of his feelings for her. Berthe Hunt walked to an overpopulated prison camp behind a barbed-wire enclosure. "What sights!" she wrote. "Some shaving, others naked, some killing cooties & others sleeping—a silence seemed to pervade the whole place." The first telephone lines went into St. Mihiel.[53]

Three days of sunshine followed three days of rain. Nighttime air alerts sounded as the moon came from behind the clouds and allowed German bombers to see targets below. First Army headquarters prepared to move to Souilly, twenty miles north, to start the promised offensive in the rocky Argonne forest.

Rumors flew around the military camp. Once again the women wondered what their role was to be. "We hear we are leaving in a few days," Hunt wrote. "We feel it is up Souilly way." She pined for her husband whenever she wasn't working. Too much time on her hands meant "a bad lonely sad day." She wished she could see Rube to say the things that could not be said in letters read by censors.[54]

During moments of calm, Chief Signal Officer Parker Hitt took meals with the operators. Two days after the victory, on September 17, Hitt was walking with General Pershing on the last sunny day when the commander crossed a bustling street to speak with three of the women. Indeed, Hitt may well have prompted the general to see for himself how the Signal Corps operators were faring. General Pershing greeted the women and asked "if they were happy they were so near the front."[55]

The atmosphere of the field headquarters was very different from the formal, hierarchical tone of the rear. There was an intimacy and directness to communications. Even so, the women must have assumed their most martial bearing. They answered, they only "wanted to be nearer."[56]

Pershing studied the operators' faces. The Meuse-Argonne Offensive would start in nine days. The fate of millions depended on the outcome. The American First Army needed whatever contribution each soldier could make. Pershing later wrote in his memoir that female telephone operators were one of the army's "crying needs" when he first recruited them. "Some doubt existed among the members of the staff as to the wisdom of this step," he admitted, "but it soon vanished as the increased efficiency of our telephone system became apparent."[57]

That clear, fateful September day in 1918, the general gazed thoughtfully at the women, then turned to Colonel Parker Hitt. Pershing's command was concise. "Take them where they want to go."[58]

Three days later, Pershing cabled the War Department for another 130 female operators immediately, and 40 more every six weeks through 1919.[59]

9

WILSON FIGHTS FOR DEMOCRACY
AT HOME

D URING THE SAME WEEK that the Hello Girls packed
their kits and headed north to Souilly, President
Woodrow Wilson geared up for battle, too.

Supporters and opponents of women's suffrage had repeatedly postponed the Senate vote on the joint resolution that
the House of Representatives approved in January 1918.
Whenever either side thought the other likely to win, they
pushed off the decision. From January through September,
Wilson doggedly campaigned with Democratic senators to
get women's suffrage across the finish line.

A master politician, Wilson had previously convinced the
legislature of matters far more controversial than votes for
women—or so it seemed. Why should this be any different?

Wilson had always moved boldly. No president since John
Adams had given a speech on Capitol Hill, yet Wilson shook
off this century-long deference to the separation of powers
during his first year in office. More than once the former university president took the lectern to advise Congress on
matters of national concern. Between 1913 and 1916, he helped
cajole the legislature into numerous measures preparing the
country for the twentieth century: the first monetary system

since the Second Bank of the United States died in 1836, stronger regulation of corporations, and path-breaking protections for workers and farmers. Exploiting the Democratic majority in both houses, Wilson's administration oversaw an epic restructuring of government. The president needed just one more domestic reform to help him restructure the world itself.[1]

If approved, women's suffrage would augment Wilson's power at the moment he needed it most. Midterm elections dominated the 1918 political agenda. Wilson believed they could spell the failure or success of a "peace without victory" once the war ended. Leading Republicans like Henry Cabot Lodge and Theodore Roosevelt made Wilson's postwar proposals a focus of the midterm campaign, calling them "silly" and "mischievous" and asking Americans to repudiate the Fourteen Points.[2] The Democrats still controlled Congress. The president hoped they would continue to do so.

The Fourteen Points reflected not merely Wilson's hopes. The war had begun over two issues: the yearning of ethnic groups under foreign domination for nation-states of their own (passions that prompted terrorist attacks like the assassination of the archduke), and the right of small countries like Belgium to protection from bigger ones. Across Europe, there were calls to establish a league of nations to address such problems. Lord James Bryce, ambassador to the United States, headed the British commission that began exploring this very possibility only a few months after the war broke out.

Historians sometimes call this the Wilsonian moment: when the idea of international cooperation to protect national self-determination took hold. Ever since, whether they approved or rued American leadership in the creation of the League of Nations, politicians and scholars have described the organization as the brainchild of Woodrow Wilson. Yet little could be further from the truth. As with women's suffrage, the United States was behind the times. The idea of an international league for peace had swirled around Europe for many generations. Americans became interested only at the start the twentieth century, having finally stabilized their own fractious federation. British, French, and even German proposals for an international league preceded Wilson's. Nonetheless, the idea gained new luster when endorsed by the large, traditionally neutral nation—and fresh urgency after four years of devastating conflict. If America championed the idea, it might actually work.[3]

This was a big "if." Victors would have to set aside their desire for vengeance. Wilson would have to secure a fundamental shift in American foreign policy, meaning a permanent political commitment to the rest of the world. His own party must remain dominant to achieve this, he believed. By 1918, fourteen states had enfranchised women, including New York and California. Those women were a massive and untested addition to the electorate. They would vote for whichever faction best served their interests. Wilson wanted to ensure they saw his constituency in this light. The Democratic Party needed women, and Wilson needed the Democratic Party.

Unfortunately, the president's party was also the biggest obstacle to women's suffrage, just as it had been to suffrage for freed slaves fifty years earlier. As a consequence, those African Americans who could vote generally followed the party of Lincoln. Wilson hoped his southern colleagues would see reason now. If he could convince Congress, women might reward the Democratic Party as their champion. Wilson had to fight his party to help it prevail.

As before, Lincoln's party led on the question of enfranchising the disenfranchised. In the January 1918 vote on suffrage, Republicans in the House of Representatives endorsed
the proposed constitutional amendment by a margin of 164
to 33. Democrats were cleanly divided, 104 in favor and 102
against. Only Republican support allowed the bipartisan
measure to obtain the requisite two-thirds majority. Nor were
Republicans shy about reminding voters. As midterm elections approached, former president Theodore Roosevelt told
national chairman Will Hays to "be very careful to see prominent women" whenever he visited a suffrage state.[4]

The Grand Old Party had not always been so supportive.
From 1897 to 1911, when Republicans controlled the executive and legislative branches, female suffrage never made it to
the floor of Congress. But the popular mood had changed, and
Republicans now sailed ahead of their rivals on the rising tide,
eager to recapture Capitol Hill and the White House with the
help of newly enfranchised voters.

Although Woodrow Wilson had once opposed women's
suffrage, he now became its most important advocate. As
American troops poured into France throughout the spring
and summer of 1918, the president repeatedly wrote key
Democratic senators whenever it seemed the suffrage bill
might come to a vote in the conservative upper house. In May,
Wilson petitioned seven die-hard opponents to reconsider. He
understood that doing so would require unusual political
courage.

As he told Senator Josiah Wolcott of Delaware, "I am
writing this letter on my own typewriter (notwithstanding a
lame hand) in order that it may be entirely confidential and
may not in the least embarrass you if you should find that you
cannot yield to this very earnest request." But the stakes were

extraordinary, he reminded the senator. "The next Congress must be controlled by our genuine dependable friends; and we may lose it—I fear we shall lose it—if we do not satisfy the opinion of the country in this matter [of suffrage] now."[5]

Wilson made the same pitch to others in his party, though they showed little interest. Senators from Kentucky, South Carolina, North Carolina, and Florida all replied that their constituents opposed women's suffrage, and they did, too. The senators cited multiple reasons, from the belief that votes were unfeminine to the conviction that any expansion of the franchise would raise the specter of black suffrage. States' rights must be defended at all costs.[6]

In June, when the amendment was again poised to come to a decision, Wilson invited suffragists Carrie Chapman Catt and Anna Howard Shaw to the White House to publicize the connection between female suffrage and foreign policy, and highlight international momentum on behalf of women. The French Union for Women's Suffrage had written Wilson earlier that year asking for his views. For the benefit of newspapers, the president asked Catt to convey his response, which she immediately released to the national press. "I welcome the opportunity," Wilson's letter read, "to say that I agree without reservation that the full and sincere democratic reconstruction of the world for which we are striving, and which we are determined to bring about at any cost, will not have been completely or adequately attained until women are admitted to the suffrage."[7]

That same month, Wilson met with a group of pro-suffrage Democratic legislators to stiffen their resolve, and wrote yet another senator who might be a swing vote. As Wilson told John Shields of Tennessee, "I feel that much of the morale of this country and of the world, and not a little of the faith which the rest of the world will repose in our sincere adherence to democratic principles, will depend upon the action which the Senate takes in this now critically important matter."[8]

When Shields replied that women's suffrage had nothing to do with the war, and that southerners could hardly vote for any expansion of the franchise, Wilson sent a second plea. Women's suffrage mattered to the Democratic Party and the world, the president argued. "I do earnestly believe that our action upon this amendment will have an important and immediate influence upon the whole atmosphere and more of the nations engaged in this war." Shields remained unconvinced.[9]

In September 1918, the week after the victory at St. Mihiel and just as Pershing began moving toward the Argonne Forest, Carrie Chapman Catt urged President Wilson to put pressure on yet another southern member of the Senate, newly inducted to take the seat of one who had died. The appointment gave supporters renewed hope of bringing the bill to a successful conclusion. The measure needed only two more votes to squeak by. Wilson met with both the governor and the senator from South Carolina, asking them not to tie his hands in the reconstruction of the world. He reiterated his plea three days later and sent telegrams to five more southern Democrats.

On the Sunday morning of the week the amendment was finally supposed to come to a vote (Tuesday, October 1, 1918), a desperate Carrie Chapman Catt wrote the president, urging him to avow even more explicitly that suffrage was necessary to win the war. Pro-suffrage senators visited Wilson at the White House with the same request. The amendment might fail. They seconded an idea that Wilson's son-in-law had suggested that very morning: address the Senate personally.[10]

A fellow southerner, William McAdoo had been Wilson's campaign manager before he married the president's youngest daughter, Eleanor, at a White House ceremony in 1913. Both Eleanor and her two older sisters supported the cause. As Jessie Wilson Sayre had written in a public letter to soldiers headed to France "to fight for Democracy" on the eve of the New York vote on women's suffrage in 1917, she hoped they would enfranchise women like her "so that we may have the full power to do our share in the great conflict fully equipped."[11]

William McAdoo urged his father-in-law to visit the Senate and tell lawmakers to do the right thing, as the president had on other occasions. Wilson vacillated. There was no precedent for addressing Congress on a pending bill the day before a vote, only on general questions or the state of the union. If Wilson pushed explicitly for suffrage as a war measure, southern senators might be able to tell their constituents that they had no choice in a matter of national security. But Wilson also risked alienating those who felt compelled to resist overt pressure or who resented any infringement on senatorial independence. The president listened patiently, "as he always did," McAdoo wrote in his memoirs, but "he thought it would hardly be possible to change their minds—nothing whatever could change them, he thought—but he promised to consider my suggestion and let me know his decision."[12]

The November elections were barely six weeks away. Militant suffragists had vowed to oppose every Democratic candidate "unless the suffrage amendment to the United States Constitution has been disposed of in the Senate elections." Across the country, Alice Paul's National Woman's Party was already campaigning against Democrats. Two weeks earlier, militant suffragists had held a boisterous protest in Lafayette Park, where they showed their disdain both for Wilson and for less confrontational activists. Shortly after Wilson issued a statement of support to a delegation led by a respectful Carrie Chapman Catt, organizers for the Woman's Party stood at the base of Lafayette's statue and torched a copy of the statement to show their "burning indignation" at the president's weak promises.[13] Wilson faced the awkward question of how to praise a cause whose rowdy proponents insulted him outside the White House.

The Senate had been debating the question for three tense days. Senators accused one another of posturing for partisan gain, putting the bill to a vote or taking it down merely to advance their interests. James Reed of Missouri fulminated at length against "discontented women" who had disgraced "the very front yard of the White House" while American boys were dying, and who had hurled "anathema and insult upon the Chief Magistrate of the Nation." Senator Key Pittman of Nevada, a suffrage state, retorted that Carrie Chapman Catt's organization had conducted its campaign in "a ladylike, modest, and intelligent way." The Senate should consider them the true representatives of woman's suffrage, not the militants—but the vivid images from Lafayette Park consumed more attention in the fractious debate.[14]

Opponents raised objections as old as the Bible: A woman's place was in the home. The ballot would destroy marriages and injure women's delicate constitutions. Those who could not physically defend the country were not entitled to vote. Adversaries' most heated objections were of Civil War vintage. Virtually every southern opponent argued that his region should not be subjected to another federal dictate on suffrage.

The Constitution of 1789 had given states the right to determine voter eligibility. The Fifteenth Amendment broke that rule in 1867, and southern states had worked ever since to mitigate its effects, passing poll taxes and grandfather clauses to disenfranchise blacks. (Their laws had the effect of disenfranchising many poor whites, too.) "I bid you pause before you put your feet into the very footprints of the same mistake that your fathers made just after the Civil War," Senator Thomas Hardwick of Georgia thundered. Even northern Republicans, he said, now admitted that black suffrage had been ruinous.[15]

Northerners and southerners barely disagreed on race policy in this era. Debate over the vote for women had replaced legislators' former anguished conversations about slavery as if the Civil War had never happened. Suffrage supporters repeatedly pointed out that race was irrelevant since neither political party wished to alter the racial status quo. Revealing the sordid political compromises of the early twentieth century, not a single senator defended the extension of the vote to black males five decades earlier. When some admitted, as Senator Hardwick did, that he feared "to the very depths of his being and the bottom of his soul the danger to the South if there is any tinkering with this question," others responded that the Nineteenth Amendment could not possibly threaten Jim Crow. Quite the opposite, they said. Women's suffrage would increase the number of white voters helping to keep blacks down.[16]

As Senator Kenneth McKellar of Tennessee told colleagues, "Any person who really wants white supremacy in the South can not better guarantee it than by the enactment of this equal-suffrage resolution." White females greatly outnumbered black females, so white voter rolls would grow. Furthermore, Jim Crow would nullify any marginal increase in black voters. Poll taxes and grandfather clauses had eliminated "the ignorant negro men vote, and they will eliminate the ignorant negro women vote," he argued.[17]

Senator Joseph Ransdell, a pecan planter from Louisiana, concurred. Confident in white dominance, Ransdell endorsed women's suffrage without qualms. "In my judgment the situation as to negro women can be handled as has been done with negro men for the past 25 years." In most states, African Americans simply did "not attempt to vote," Ransdell said, phrasing his point in such a way as to omit the well-known reason blacks did not show up, namely to avoid death by lynching. "It is inconceivable that these conditions will be destroyed or even interfered with by permitting women to vote."[18]

James Vardaman of Mississippi, another legislator who supported women's suffrage, was more racist yet. He admitted they risked enfranchising "a few negro women" who would be even "more offensive, more difficult to handle at the polls than the negro man, for 'verily the female of that species is more deadly than the male.'" And when the war ended, "the arrogance and impudence of the ex-negro soldier will greatly enhance the white man's burden." But this was precisely why white women should get the vote—to help southerners carry their peculiar cross. Vardaman announced his intention to fight for the repeal of the Fifteenth Amendment to eliminate all "negroes, both male and female, from the politics of America"—as well as his conviction that "every normal white woman in the South" would help if only she could vote.[19]

Feminists made little effort to disabuse racist congressmen of their convictions. No amendment could succeed without them. As historians observe with regard to Franklin Roosevelt's World War II partnership with Stalin to defeat Hitler, untainted allies were not an option.[20] Yet even the support of

Vardaman, Ransdell, and McKellar did not suffice. The majority of southern senators could not bring themselves to endorse a measure that enhanced federal power over a matter that touched on old scars—even if it delivered them new segregationist voters.

This contradiction is worth pondering. The repugnant debates around race may obscure a deeper reason for prolonged foot-dragging on women's suffrage: the fear of revising gender roles so ancient as to have no date of origin. While many southerners fell back on the self-pitying claim that northerners just didn't understand their "race problem," not a single northern senator openly disagreed about disenfranchising blacks. The race card thus served as a form of misdirection, like a magician's trick. It was a convenient excuse that allowed opponents to deny females the vote without admitting they simply could not imagine women as political equals.

Senator William Kirby of Arkansas, a supporter of suffrage, openly asserted this. White supremacy was unchallenged, he pointed out. Yet other "epochal" changes were sweeping the world, and men must learn to adjust, no matter how hard it seemed. "I suppose I was 21 years old before I ever saw a woman riding horseback astride," Kirby pointed out, "and I have not gotten used to it yet." Such unladylike behavior was simply not permitted where he came from. But as he matured and observed women in other sections of the country riding "astride, cross saddle," he realized it was simply safe practice. The prejudices of childhood remained rooted in his heart, but he knew decisions must be based on adult reasoning. "We have been bound by habits of our people, by our environment; and some of us have not been able to look beyond," Kirby stated. "That, I believe, is the trouble with some here today."[21]

Senator Porter McCumber of North Dakota expressed a similar unease. He intended to vote for women's suffrage because his state had passed a piecemeal plan of limited rights that he felt compelled to honor. But McCumber nonetheless questioned the "wisdom and propriety." Legislatures across the land were succumbing to feminists' demands despite men's instinctive recognition, "way back in the masculine consciousness," that the vote might "shatter the social structure builded [sic] through centuries of human progress and so admirably conforming to the different powers and functions of each sex."[22]

From the "time of the caveman"—who brought meat to his woman, who fed his "little brood safe behind his sheltering arm"—separate sex roles had defined society. These differences fostered interdependence. McCumber would endorse the bill, but he feared the long-term consequences of men and women sharing the same jobs. He also noted that women were not equally qualified to exercise judgment in matters of national security, raising a subject that came up repeatedly during the debate, when news of bloodshed in Europe grabbed headlines daily. "The soldier's sword could never be fashioned to fit her hands," he warned the Senate. A woman's sensitive nature made her prey to pacifism. Although McCumber did not name Congresswoman Jeannette Rankin, other senators undoubtedly recalled her voting record when McCumber warned that female politicians would never approve war.[23]

At the time, women were prominent, even dominant, in pacifist organizations.[24] Jane Addams was president of the Women's Peace Party. Bertha von Suttner, the Austrian leader of the international arbitration movement, became the first female recipient of the Nobel Peace Prize in 1905. Some British suffragists attended a peace congress in The Hague in 1915, while more nationalistic suffragists condemned them. Quaker Alice Paul privately advised Rankin to vote against Wilson's declaration of war. Historian Roland Marchand observes that many suffragists equated the female franchise with world peace, since women's "moral sensitivities" made them more pacifistic than males. Senator McCumber agreed. In dangerous times, he claimed, America needed "the masculine hand and the masculine heart, with all its unyielding firmness—yes, with all its belligerency—at the helm."[25]

From the late eighteenth to the early twentieth centuries, new nations like America and Germany, and older ones like France and Britain, fused patriotism, masculinity, and war. The rise of nationalism clinched the association of manliness with soldiering, previously a limited, specialized occupation. Universal conscription accompanied the spread of nation-states. For the first time in history, a broad spectrum of European males faced the draft. In popular imagination, real men became heroic, real women damsels in distress. Violence was male, nonviolence female. Soldiering for one's country was thought necessary to forge patriotic men.[26]

Yet the European war provided evidence that challenged these assumptions. Some men called for peace without victory; some women embraced responsibility for defense. The most compelling arguments in favor of suffrage sprang from actual events, which disproved hypothetical objections to females as inadequate, lesser citizens. Women soldiers, sailors, and marines were helping win the war. Nearly every belligerent had already recognized female contributions by granting the ballot. The United States was damaging its credibility as a democratic leader, supporters alleged.

John Shafroth of Colorado, a first-term Democratic senator whose seat was in jeopardy heading into the midterm election, pointed out that while some people believed that women "should not have the right to vote because they could not fight for their country," modern warfare nullified this objection. An "army of women war workers" labored across the country and in Washington itself. In France, American nurses kept American soldiers alive. Shafroth did not single out telephone operators, but he praised female volunteers driving ambulances and trucks near the front. Even if women did not bear arms, he said, "in the balance of their country's distress . . . they have not been found wanting." Veterans would give women the vote once they returned if the Senate did not do so immediately. "I know they realize the splendid services of these women who have been sent to our battle fields."[27]

Shafroth regaled the Senate with the story of Russia's "Battalion of Death," the famous women's brigade that had fought until the Bolsheviks signed the disgraceful Treaty of Brest-Litovsk. "There is no sacrifice that woman will not make for her country," Shafroth claimed. If America should need its women to perform comparable feats, they would do so. "When the absolute necessity for physical force would come, you would find that women, though not so strong as men," Shafroth argued, "would give that determination to their action which would produce a defense beyond calculation." Shafroth challenged his Democratic colleagues: "[How] can it be asserted that women are not citizens when they perform almost every task that man has done?"[28]

William Thompson of Kansas pleaded for the amendment as well. Like Shafroth, he was another first-term senator from a state where women could vote—and had been urged to eject Democratic incumbents. "Women have performed more than their part in this great struggle for democracy, freedom, and liberty," Thompson said. In the United States, France, and

England, women were producing food, guns, ammunition, planes, trains, and every other necessary item. They loaded baggage, drove trucks, and were "ready, if necessary, to shoulder the gun and march to the front themselves," he claimed, perhaps in reference to British WAACs, American marinettes, Signal Corps operators, or even Russians. Thompson had traveled in the war zone, where he had "never heard anyone even hint that all this work was not being properly and efficiently performed." Rather, women were praised everywhere.[29]

Senator Ransdell of Arkansas pointed to the irony of democratic America failing to enfranchise qualified adults. "Women in several of our States and in many countries of the world have voted for years, and there is no evidence that they are less womanly than before," he said. "Is it possible that this great, free country, which is spending countless billions of wealth and giving thousands of its best young lives to 'make the world safe for democracy' is going to longer deny this plain, simple act of justice to the better half of its citizens?" In short, the United States must practice what it preached to others.[30]

Ransdell's defense of women's rights falls strangely on modern ears given that he simultaneously applauded Jim Crow. As contradictory as this was, Ransdell spoke in a context where, internationally, minority rights commanded little attention. Although there was growing acceptance that substantial ethnic clusters within large empires should have their own national borders, such as the Irish in Great Britain or the Poles in Central Europe, there was zero appreciation for what is today called diversity and little consciousness that subminorities deserved protection. "Subject races" generally meant Czechs or Lithuanians. In contrast, women around the world were winning the vote, reminding Americans that they were tardy outliers to a trend others considered important.

Although W. E. B. Du Bois called the problem of the twentieth century "the problem of the color line," discrimination on the basis of skin tone did not attract wide condemnation until after World War II. (When that happened, opponents harnessed it to pull down Jim Crow at last, aided again by presidents embarrassed at world opinion.)[31] In comparison, disgust at the failure of democracies to enfranchise female adults reached a tipping point around 1918.

As Republican Wesley Jones of Washington lectured the Senate, "You can no more stop it than you can reverse the decree of Sinai." Times had changed. Women's suffrage was the "new order of things," and America "must meet this in an up-to-date way." Jones quoted English prime minister David Lloyd George to bolster his case. British women had made the ammunition that saved the army on the Somme. It would be "an outrage," the English politician told his people, to exclude them from government any longer. "That is why the woman question has become very largely a war question."[32]

Woodrow Wilson agreed. While Senator Jones was speaking to the full chamber, the president drove down Pennsylvania Avenue to the Capitol. He had decided to address Congress. One motivation may have been the shocking announcement of Christie Benet, the potential swing vote from South Carolina, that he opposed the amendment on the grounds of states' rights—regardless of the consequences for the midterm elections. "This is a matter of principle which rises high above any party interest or possible party defeat," Benet proclaimed. Southern women did not "expect or want any reward" for their patriotism.[33]

Wilson's daughters and new wife watched from the gallery as he took the podium amid legislators hurrying to their seats.

The president's appearance "was entirely unexpected," the *New York Times* reported. William McAdoo had labored to keep word from getting out ahead of time. Legislators leaned forward or settled back, listening intently while Wilson read a short prepared speech. Both suffragists and antisuffragists peered down anxiously from the visitors' balcony.[34]

Tension filled the chamber. "It was clear that his appearance was bitterly resented by all those opposed to the amendment," McAdoo observed ruefully. Even proponents appeared miffed by Wilson's interference. Yet "an air of hostility, a frigid atmosphere," tended to rouse the president's oratorical powers, McAdoo thought. Wilson's speech was "powerful and impressive and carried a fighting edge."[35]

He reminded the Senate that the "unusual circumstances of a world war" meant that not only fellow citizens but also "all nations and peoples" were watching. Wilson's next sentence was blunt. Women's suffrage was "vitally essential" to the winning of the war. America must listen to the "plain, struggling, workaday folk" around the world, looking "to the great, powerful, famous Democracy of the West to lead them to the new day for which they have so long waited." Alluding to female protesters in Lafayette Park, Wilson dismissed the notion that "foolish and intemperate agitators" had influenced him.

The United States simply could not hold itself aloof from world opinion if it wished to lead. Other countries had expanded the franchise or were poised to do so. "Are we alone to refuse to learn the lesson?" the president asked. Was the trust and respect of other free nations "an asset" or not? The United States must face such questions. Other liberal peoples had accepted a new interpretation of democracy. "We cannot isolate our thought or our action in such a matter from the thought of the rest of the world," he proclaimed. "We must either conform or ... resign the leadership of liberal minds to others."

The president focused on female citizens who had done their simple, plain duty, some "upon the very skirts and edges of the battle itself." The war could not be fought without them. "Are we alone to ask and take the utmost women can give—service and sacrifice of every kind—and still say that we do not see what title that gives them?" the president asked. "Shall we admit them only to a partnership of sacrifice and suffering and toil and not to a partnership of privilege and of right?"

Wilson spoke only fifteen minutes, but his words rang boldly in the hushed chamber. No one could have defended women more eloquently or linked their rights more directly to the Great War. Wilson reminded listeners of his heavy responsibilities. "I ask that you lighten them," he pleaded, "and place in my hands instruments, spiritual instruments, which I do not now possess, which I sorely need, and which I daily have to apologize for not being able to employ."[36]

Wilson's endorsement thrilled Carrie Chapman Catt, who told newspapers, "No country has paid so great a tribute to its women as was paid the women of our country today."[37]

But the president had not converted a single, solitary senator. He had hardly left the chamber when Oscar Underwood of Alabama, normally one of Wilson's strongest supporters, jumped to his feet to denounce any federal amendment as mob rule. John Williams of Mississippi inveighed that he loved the president like a brother, and did not wish to increase the man's burden by a feather, but could hardly take seriously the claim "that we cannot whip Hindenburg, that we cannot outmaneuver Ludendorff, that we cannot scatter the Bulgarians ... unless the negro women in the State of Mississippi can vote."[38] When roll was taken, Democratic opposition went far to defeat the joint resolution. Of the twenty-two senators who refused to help their party leader, nineteen were southerners.

A handful of Republicans also voted against, including conservative Henry Cabot Lodge, a steadfast opponent of female suffrage. Another was George McLean of Connecticut, a former Harvard classmate of Teddy Roosevelt, who reminded the Senate that Jeannette Rankin had demonstrated precisely the soft material of which women were made. When subjected to the "acid test," Rankin had "proved herself to be just what God made her—a womanly woman." The Almighty may have erred in making males stronger, McLean said, but "all questions involving declarations of war and terms of peace should be left to that sex which must do the fighting and the dying on the battle field."[39]

Racial prejudice dressed up as states' rights, combined with the belief that women could not physically defend their country, had defeated the Susan B. Anthony Amendment. Wilson expressed disappointment to his Mississippi colleague not long afterward. "I must say frankly that I was very much grieved that the Senate did not respond to my appeal about woman suffrage the other day," he wrote John Williams, "because I knew what I was talking about when I spoke of the effect it would have . . . and the effect is going to be very serious."[40]

The next month, a week before the war ended, Republicans gained forty-four seats in the lower chamber in the November 5 elections, and a majority of one in the upper. The Democratic Party lost leadership of both the House and the Senate. Among those who forfeited their seats were Senators John Shafroth of Colorado and William Thompson of Kansas. Farmers' discontent over falling wheat prices hastened their defeat, but suffragists' opposition did not help. "We ask the voters to remember that the Democratic Party defeated suffrage in the Senate, in spite of the President's appeal to pass it as a war measure," Alice Paul proclaimed before the election. "Since then the Democratic Party has done nothing to win the two votes still lacking." In her view, inaction had to be punished.[41]

But while the federal amendment suffered a setback, women in states where suffrage had recently passed took delight in newfound powers. In New York City, they showed up earlier than men at the polls and parked baby carriages out front. The governor and mayor posed for pictures with their wives marking ballots for the first time. Carrie Chapman Catt cagily told the *New York Times* that she had split her vote between Democrats and Republicans. Another suffrage leader informed reporters, "It seemed as natural as breathing, and I felt as though I had always voted."[42]

Woodrow Wilson's loss was more permanent. Republicans handed the chairmanship of the Senate Foreign Relations Committee to Henry Cabot Lodge, who despised the president. Had Democrats retained even a couple seats, the consequences for the postwar world would have been different. Lodge used his position to ambush Wilson's foreign policy.[43] Defeat of women's suffrage helped defeat U.S. participation in the League of Nations.

The vote and the president's peace plans had gotten snagged on the thorniest problem of the American republic: racial prejudice. A Virginian had championed a measure enfranchising black women, yet even he could not steer such a bill through a Democratic Congress.

If Grace Banker still resided in New Jersey, where women had yet to win the vote, she might have watched through the windows as female New Yorkers pushed their babies past AT&T to polls in downtown Manhattan. But war in France allowed her to practice citizenship in yet another way.

10

TOGETHER IN THE CRISIS OF
MEUSE-ARGONNE

H OW FAR could sex roles stretch without men becoming "unmanly" or women "unwomanly"—as then
defined? How dependable and useful would female soldiers
prove in a supreme crisis? Could men and women rely on one
another implicitly and without awkwardness in close
quarters?

The Great War raised such profound questions because
nationalism stirred women as greatly as men and because
industrialization created new demands for female labor.
Congress debated these questions in hypothetical terms.
At the Battle of Meuse-Argonne, an army of men and a
handful of women attempted to answer them in practical
terms.

No one knew that this would be the largest, longest, deadliest single battle in all American history; that it would kill
more than 26,000 and injure another 95,000 in the course
of forty-seven days. But commanders on both sides understood that the fall offensive might decide the war. In it,
America would prove what it could bring to the struggle, and
so would the Hello Girls.

Telephone operator Berthe Hunt sensed the scale of the looming contest as they passed signs for Verdun. Devastated lands and a concentration of weaponry made plain they were following Pershing into hell. "As we neared Souilly we began to see houses that were in ruins, barracks camouflaged . . . Ammunition dumps with the big mounds of dirt . . . [and] miles & miles of trucks & artillery."[1]

Hunt was aware that she and other operators could be among the casualties. "Big guns going off," the Californian noted her first morning on duty. "We hear that we will probably be bombed tonight," she wrote casually two days later. "I'll go to bed dressed in case it comes before 11 P.M.," the hour that Hunt needed to report for duty.[2]

No bombs fell, but the next afternoon, September 24, U.S. antiaircraft guns brought down a German plane and dropped heavy shrapnel four feet from Grace Banker, who had ducked outside the office for a breath of fresh air in the rare sunshine of autumnal northern France. As she gazed up at the enemy sailing in the blue overhead, the aircraft looked as harmless as a dragonfly, and the shot of the guns like bolls of cotton, until "the shrapnel which had gone up began to come down." For the first time, Banker faced reprimand. Horrified to learn that his chief operator had gone outdoors without a helmet, Colonel Parker Hitt scolded Banker sternly. He might have lost her.[3]

Hard work accompanied such dangers. The telephone lines were always heaviest right before a drive, as troops struggled to get into position and supply units rushed to equip them. Switchboards burned continuously. Colonel Hitt kept the operators on twelve-hour shifts around the clock, still not trusting males to handle the overnight traffic. The First Army faced its most important battle only a few days after the last. It was a logistical nightmare, for which Pershing had unwisely volunteered.[4]

Black Jack planned to surprise the Germans. By day, artillery convoys melted into the forest. By night, the army moved forward. A colonel in the AEF medical corps described the nocturnal advance as "prowling, skulking, preparing, stalking, 500,000 armed human beings accompanied by acres of guns—paraphernalia covering the earth—a blanket of destruction ten miles deep, thirty miles long, gliding by inches, skulking by inches—hundreds of thousands of my fellow beings are dragging and lugging this vast carpet of destruction towards the enemy."[5]

At the center of this activity was the telephone contingent, a microcosm of all women in uniform. Colonel Parker Hitt felt confirmed in his advocacy of them. "There were a number of times when the load . . . passed belief," he reported to superiors. "On a number of occasions all 75 of the cords were up and drops still falling."[6] In the wee hours preceding the great offensive, Berthe Hunt connected calls without ceasing to the backbeat of a heavy barrage that the French Army opened to deflect attention from American maneuvers. German guns boomed in reply.

The cannons woke General Pershing around 5:00 A.M. on September 25. He used the opportunity to make calls from his private train, again partially hidden in the nearby woods. Berthe Hunt helped the commander reach those he needed. The artillery barrage was so close that Pershing suddenly yelped in her ear. "He let out an exclamation & told me a shell had just whizzed past his window."[7]

Finally, the great offensive began just past midnight. In the dark skies above the battlefield on September 26, America's ace pilot Eddie Rickenbacker surveyed the opening bombardment with glee. The Allies appeared to be launching a million guns simultaneously. The panorama below made him "think of a giant switchboard which emitted thousands of electric flashes as invisible hands manipulated the plugs," Rickenbacker wrote. One of the officers commanding those guns was Captain Harry Truman from Independence, Missouri.[8]

At the actual switchboards, Grace Banker stayed on duty twenty-one hours that day. "Don't believe anyone not here can realize what busy means," she jotted later. Banker ventured momentarily into the icy dark when the barrage began, accompanied by an army captain and Julia Russel, the fearless YWCA secretary in charge of billeting who refused to be parted from the women's contingent. Field cannons roared in their ears as they scanned the darkness. Banker saw "great flashes of light all along the horizon like the Northern Lights." Presumably, she wore her trench helmet this time.[9]

The unbelievable noise became even louder around 3:00 A.M., when the big railroad guns commenced firing. Somewhat more than two hours later, at 5:30 A.M., a quarter-million men swarmed toward the German lines. Throughout that day, and many of the following forty-six, the flimsy wooden walls of the Signal Corps barracks shook, "and the beds rock[ed] as though in a miniature earthquake."[10]

Situated next to a bomb shelter that contained an underground switchboard for emergencies, the billets were their most Spartan accommodations yet. Sleeping quarters lay at one end of a long, ramshackle, tar-papered shed used by French troops during the tragic Battle of Verdun.

Number 8 in a row of barracks, the wooden building floated on a sea of mud at the edge of shell-torn Souilly, the last inhabited village before the Argonne Forest. The women shoved newspapers and old maps into cracks to keep out the cold. The roof leaked so badly that Grace Banker had to move her cot repeatedly to find a dry spot. Window frames held no glass, just oiled paper, and looked like the hatches of a chicken coop. At night, the women hooked them closed and pulled a black curtain across the paper. If light escaped, a sentry banged on the wall with a gun and roared at them to draw their camouflage more tightly.

Signal Corps officers worked at the opposite end of the building in offices christened the "sacred precincts" by the operators. Between them was a small common mess, where a dozen or two officers and operators came together for meals at a long wooden table made up of sawhorses and pine planks. Benches flanked the table. Manners relaxed, and the women learned to eat like men. "You use your own fork to help yourself to the bacon," Banker observed with delight. She began drinking her coffee black. "As for putting away hotcakes in the morning and hash and beans at night I am sure I have no equal."[11]

Colonel Parker Hitt commanded the head of the table, Suzanne Prevot often on his right, with women interspersed among the men. Not every operator took immediately to the egalitarian spirit. One wrote home, "I was considerably abashed when I set down for my first meal here with a long row of officers on every side. Perhaps you cannot imagine how it seems to be flanked on the front and sides by Colonels, Majors, Captains, and Lieutenants."[12]

Two African American soldiers served the rations, dished onto tin enamelware by a Louisiana cook whose accent, one operator wrote, was "so peculiar that I took him for a foreigner."[13] She betrayed the segregated army's racial mores when she noted that the friendly orderlies "are very amusing and will do the darky jig for you at any time." The only African American combat unit, the Ninety-Third Division, was assigned to French commanders used to colonial troops. The French readily accepted such assistance, unlike prejudicial British officers who declined Pershing's offer of another such unit, the Ninety-Second. From 1917 through 1918, Black Jack struggled to employ Buffalo Soldiers without fighting a pyrrhic campaign against racial bigotry at the same time. At Meuse-Argonne, the only black combat troops wielded French rifles. An estimated fifteen hundred African American nurses certified by the Red Cross were also barred from service until the influenza pandemic persuaded Washington to induct eighteen pioneers on an experimental basis.[14]

Adjoining the Signal Corps mess was the exchange, a large bare room with a small stove for warmth and the portable switchboard that operators staffed continuously. Each was responsible for fifty lines, all connected to G3, the U.S. Army's Operations Section. As Berthe Hunt later recalled, "Every order for an infantry advance, a barrage preparatory to the taking of a new objective and, in fact, for every troop movement, came over the 'fighting lines,' as we called them." On a typical night in early October, amid hundreds of calls, she took one from a Signal Corps office under bombardment, another from a French operator trying to alert Americans to German planes overhead, and another from an artillery division that needed to know the precise time in order to start firing when commanded. "I worked hard to get *exact* time to the second—a push is coming off tonight again!" Hunt wrote in her diary.[15]

Beyond the exchange, separated only by a flimsy partition, six operators slept in shared cubicles on cots lined with straw. Summoned to headquarters the week after the battle started, Adele Hoppock made a seventh. She wrote her parents that she and her roommate were never out of hearing of the switchboard on the other side of the wall. "We go to bed and get up in the morning to the tune of, 'Are you waiting, are you through?' 'Number please,' 's'il vous plait,' 'allo, allo' and so forth.'"[16]

Hoppock's eagerness to work near the front coincided with a pressing need. Marie Lange, Berthe Hunt's roommate, fell dangerously ill with the flu a few days after the battle began. Six operators meant three women for each twelve-hour shift. "Haven't enough girls for a substitute so am going to take the shift tonight," Grace Banker noted in her diary. With only an hour's sleep in yet another twenty-four-hour period, Banker struggled to stay awake, despite drenching rain and high winds that pounded the walls.[17]

Banker did not pity herself, however. "I keep thinking of our boys in the trenches and the wounded for whom there is not sufficient transportation," she wrote. Muddy, clogged roads impeded rolling kitchens from reaching hungry soldiers. Ambulances returning from the battle faced traffic delays of twenty-four to thirty hours, their patients often dying before reaching a hospital.[18]

Berthe Hunt had similar thoughts. "Boys not making the progress planned—much resistance, everyone blue." Hunt felt discouraged, too. Yet she knew soldiers outdoors had it worse, and she spent the eve of her thirty-fourth birthday keeping up their spirits. "Talked tonight over the wires to boys in the dugouts," she wrote on September 28, two days into the offensive.

Berthe Hunt took pieces of Marie Lange's shift as well. Adele Hoppock finally arrived, making the staff shortage less acute. Berthe Hunt breathed a sigh of relief: "Hoppie come from Nch [Neufchateau] to join our ranks." Grace Banker was thankful, too. "She is a good sport and I am glad to have her."[19]

Marie Lange's recovery and Adele Hoppock's transfer eased the burden only a trifle. Like other soldiers in the advance zone, operators had no days off. Leisure meant an afternoon walk past hospitals and prison camps to fields where the dead lay buried under rows of temporary crosses, or an occasional ride to another part of the battlefield. "Cemeteries everywhere," Berthe Hunt wrote sadly. Stiff German resistance made clear that the women would be at Souilly longer than anyone had anticipated. The war was not going well. "It makes you sick to think of this struggle going on and on and so many boys dying," Grace Banker wrote on October 2. "The ambulances go slowly by bearing their burden of wounded." Colonel Hitt ordered the women to scout out whatever "necessary things" they would need for a long, cold winter.[20]

Grace Banker waited vainly for a letter from her brother, Eugene, who had not written in months. She worried he was dead but purchased cigarettes and sent them to his artillery unit, hoping he might receive them yet. She could not get him out of her mind, especially whenever she saw ambulances. After she lost a filling in a tooth, Banker's commanding officer ordered her to the nearby field hospital, where she happened to witness a surgeon digging for shrapnel in the chest of an injured soldier as casually as if cleaning a chicken. Another shell-shocked patient studied the bandaged stump of his amputated wrist. Banker felt sick.[21]

Worn through, the chief operator feared she might contract the same Spanish influenza as Marie Lange. Banker did not know, as the AEF did, that mortality from influenza-related pneumonia approached 32 percent that week. The next, it climbed to 45 percent. In some units, the death rate of those who caught the disease reached 80 percent. Stomachaches, chills, and fever beset Banker the first days of the offensive, though she stayed at her post and dreamed of quiet New Jersey. "When I get home I am going to sleep for a week without stopping."[22]

Banker and her crew struggled to maintain a soldierly appearance. Ice formed in the pails they used for washing. "We wear our white blouses as long as possible on the right side and then if we can't do any better we wear them wrong side out," Banker wrote. Their skin grew chapped, and faces lean. More than ever they felt like soldiers. When someone sent a box of candies marked "Telephone Ladies," the civilian moniker felt strangely unfamiliar. "Sounds funny for we are never called anything except the Signal Corps Girls," Banker wrote.[23]

In modern parlance, the word "girls" might sound belittling, but in 1918 it resonated differently. Male soldiers were often called boys, as in "doughboys" and "our Signal Corps boys." When officers spoke of their boys or girls, they communicated fellowship, caring, and team effort. It was praise, not put-down. Grace Banker's definition of herself as a Signal Corps girl revealed her sense of belonging to a seamless military unit with common objectives: all for one and one for all.

At Meuse-Argonne, their most immediate goal was to clear the bleak, forbidding forest of Germans barricaded in cement-reinforced trenches, pillboxes, and machine gun nests ringed with deep belts of barbed wire. Once accomplished, the army planned to press east, crossing the Meuse River to eject enemy troops from the ridge on the other side, and then push into the Woëvre plain of Lorraine. Unlike at St. Mihiel, however, the Germans had no intention of pulling back. They were fighting for survival against massive English, French, and American armies across a broad swath of territory on the approaches to the fatherland itself.

Of these armies, the Americans were the least well trained, equipped, or organized. Some soldiers had just crossed the Atlantic, never having fired a rifle or detonated a grenade. Even those trained with the rifle were hardly adept. As backwoodsman Alvin York said, his comrades "missed everything except the sky." Most American soldiers could not tell gas from gun smoke. In comparison, battle-hardened French and British troops could distinguish one poison gas from another by scent. Men who had fought at St. Mihiel had more experience than the newest doughboys but had not witnessed the German soldier at his best—or even "his second best"—in the estimation of Major General Hunter Liggett, Pershing's most knowledgeable field commander. The Meuse-Argonne was the AEF's first sustained battle against seasoned Germans.[24]

More than a million U.S. soldiers fought there. Historians have devoted entire books to the failures, setbacks, and ultimate triumph of the American Expeditionary Forces in that complex slugfest. The candid summary of Pershing's logistics expert at Souilly, Colonel George C. Marshall, serves well. On the American side of the conflict, officers struggled with "distressingly heavy casualties, disorganized and only partially trained troops, supply problems of every character due to the devastated zone so rapidly crossed, inclement and cold weather, as well as stubborn resistance by the enemy on one of the strongest positions on the Western Front." American inexperience was a profound handicap. Five of the nine divisions that jumped off the first morning, on September 26, 1918, had never seen combat.[25]

U.S. officers also made many of the same mistakes previously committed by European commanders, including frontal attacks across open terrain laced with barbed wire. "Groups of our lads dashed up to the wire only to be shot down to the last man," an observer wrote. Fallen bodies hung from fences. Field engineers could not cut down obstacles fast enough. Gas attacks caused almost 19,000 casualties, or 20 percent of those injured at the Meuse-Argonne. The Germans fought tenaciously.[26]

American forces also suffered from self-inflicted wounds, mainly lack of coordination. Pershing was overextended as both commander in chief (with responsibility for operations across France) and First Army commander (with tactical responsibility in one theater). Supply lines were too short, and soldiers did not have sufficient food, ammunition, or even water. Fog sometimes shielded troops from German machine gunners, but other times caused them to lose their way in the unfamiliar terrain. On the second day of battle, the First Army captured the strategic high ground of Montfaucon (Falcon Mountain), an infamous German lookout. The French command thought it could not be conquered before Christmas and was deeply admiring. After this, American advances largely stalled. Pershing had to draw back several times.

One of the most iconic failures was the entrapment of a group of 679 soldiers known as the Lost Battalion. The men were told to leave rain gear, tents, spare rations, and blankets behind in order to overwhelm German positions quickly. The men advanced on October 2. They bravely picked their way forward, discovering concealed machine guns only after soldiers began dropping yet nonetheless made steady headway. Unfortunately, other Allied troops failed to come up alongside as planned. No one told them that French troops on their flank had not launched. The battalion broke through the German line into a pocket in the hills, but no one followed.

They had moved too far ahead. Germans closed in behind, stringing barbed wire across the only escape route. The doughboys were cut off with minimal ammunition and virtually no food and water. Carrier pigeons were their only means of communication. The infantrymen hid in muddy shell holes filled with rainwater, pinned down by machine gun fire.[27]

When the enemy ordered them to surrender, the Americans yelled back, "Go to hell." They dug latrines at the back of the pocket, conducted nighttime forays against the Germans, held off five attempts to rush them, and hung on, praying for relief. The only source of clean water was within range of German guns. Enemy soldiers gradually decimated them. Other troops were injured by American artillery until the commander managed to send one of his last pigeons with the message "Our own artillery is dropping a barrage directly on us. For heaven's sake, stop it." After five anguished days, another American unit broke through to the besieged men, whose plight had become the talk of the AEF. Of 679 soldiers, only 252 survived. Berthe Hunt finally wrote in her diary on October 7, "Had news tonight that an American battalion that has been surrounded by Germans was rescued."[28]

Despite these and other grim tidings, Hunt and the other operators never flagged. Not that they had any choice. They were obligated for the duration. But they were also highly motivated. They considered themselves trailblazers and felt the honor keenly. The operators whom Adele Hoppock had left behind in Neufchateau were "wild to go, and they consider me very, very fortunate."[29]

Stars and Stripes confirmed Hoppock's impression. Reprinting a story from the soldiers' newspaper, the *Lexington (KY) Leader* told readers, "It was hard to pick out the ones who were to go, so anxious was the whole force to get a crack at the big show . . . [despite] all the discomforts and dangers that come with being billeted in the forward area over which the Boche avions fly when they can." The article acknowledged that some doughboys in more prosaic assignments resented the women with Pershing, especially men "toiling away further down the line and answering the call of the telephone sextet with a green and rankling envy."[30]

Yet the work also fostered respect and comradeship. The women's bond with one another deepened. Grace Banker wrote on October 8, "I like the work, and my girls are wonderful." Men also made them feel valued, from doughboys on the road to their commander in chief. "Went downtown, saw Pershing," Berthe Hunt wrote on October 13. "He saluted us— he is most democratic." Indeed, the uniformed women were saluted almost universally and cheered by the worn, battered troops they encountered.[31]

The Hello Girls knew the army counted on them. Their work was important. When operations gravitated away from Neufchateau, Adele Hoppock had been uncertain of her usefulness. "Now I know we are much needed," she wrote her parents from Souilly. "The work is hard and strenuous."[32] On the day Pershing greeted Berthe Hunt in the street, Hunt was doubly elated. Her husband had written. The threat at sea was lessening. "It seems now that he will be spared me." Hunt knew she was helping to shorten the war and save lives.[33]

Yet the women also had no idea how much longer the cruel conflict would last, and thus they worked to buoy one another's spirits. An officer brought back a piano he found in one of the more elaborate German fortifications. The women ensconced the instrument in a cubbyhole they dubbed "the parlor," and gathered with officers to play, sing, and talk during free moments. They observed every birthday, made more precious by the slaughter. Colonel Parker Hitt gave a dinner in Ligny for Suzanne Prevot when she turned twenty. Grace Banker made fudge for Berthe Hunt after they arrived at Souilly. "Girls brought flowers & we had a pretty dinner table with candles & candy & flowers so it was a birthday after all," Hunt wrote, lifted from the sadness she had felt earlier that morning.[34]

A month later, when four members of the Signal Corps mess had birthdays, Berthe Hunt planned an ambitious menu. One of the celebrants was Major Bruce Wedgewood, second in command, who sent to Paris two hundred kilometers away for the rare comestibles she suggested. On the day before her fete, Mrs. Hunt scoured the tiny country village for someone to cook the special Sunday dinner and begged flowers— "beautiful yellow marguerites"—from the French barracks. The chef she recruited from a nearby French airfield performed miracles: a gala banquet with lobster, caviar, cheese, champagne, cake, nuts, roast goose with stuffing and mushroom sauce, roasted potatoes, and fresh cauliflower. Adorning the sumptuous table was the first fresh fruit the Signal Corps had seen in what felt like forever.[35]

Electricians rigged a domed canopy of lights and flags over the old plank table. Hunt wove fernlike celery into the improvised chandelier and used the basket of yellow flowers as a centerpiece. Colonel Parker lent red and white semaphore banners, which Hunt draped along the walls. "The Signal Corps wash on the line," Adele Hoppock dubbed the wigwag flags. Grace Banker, one of the celebrants, couldn't believe her eyes. She pronounced the table *magnifique*. "Such anticipation— such excitement—such sniffing—such cooking," she wrote.[36]

The men and women exchanged riddles into the night and teased one another affectionately. The operators wrote anonymous place cards for their guests. Grace Banker penned a mischievous jingle on the card for Colonel Sosthenes Behn. Behn started International Telephone and Telegraph (IT&T) after the war, but in 1918 he was a field commander renowned for biting commentary and stern discipline. "Wit like his would surely freeze / Even grim old Socrates," Grace wrote, hoping Behn could laugh at himself, which he apparently did. Others chuckled at the joke they discovered the French barracks had played on them. Yellow flowers, it turned out, stood for flirtation and infidelity.

An informal photograph of the event reveals an evening of easy camaraderie. Despite everything, they had made it thus far. Male soldiers must have pitched in to staff the telephone exchange during the dinner, although Marie Lange is not visible in the photo and may have remained at the switchboard.

The following Friday, on Grace Banker's actual birthday, operators revealed a special affection for their leader. Tootsie Fresnel and Suzanne Prevot chipped in for a lacy French brassiere for Banker to wear under her severe uniform. Marie Lange conjured a rare spray of fall chrysanthemums. Berthe Hunt gave her "a tray of fruit of which I am so fond." The touching gifts delighted Banker. "My girls are so thoughtful."[37]

The operators also took every opportunity to laugh. Cold, overworked, and sad at the continual devastation, the women found humor wherever they could. Thick, oozy, slippery, sticky mud that came up over their shoes was a special enemy and thus a particular source of merriment. "Woe to him who falls flat on his back in the miserable stuff," Grace Banker wrote. She soon did, of course. "Fell flat on my back in the mud in front of the Artillery Barracks this morning, and got literally plastered with mud," she noted in early November. "We took a knife and tried to slice it off." Berthe Hunt "went full length" in the street four days later.[38]

Yet the women also played in it. When Hunt and her roommate, Marie, had a chance to bathe for the first time in days, they threw modesty aside—"rude, but it did feel good"—and shared a shower rigged up in an old shed. They got dirty again in the process of drying, but instead of becoming irritated, they slid around on the muddy floor and "laughed ourselves sick."[39]

Finally, in mid-October, the tide of the war turned. The first signs appeared that the enemy was weakening.

Pershing had revitalized his forces by delegating leadership of the First Army to Major General Hunter Liggett on October 12. Liggett had commanded troops in Cuba and the Philippines, and earned Pershing's trust at the Battles of Château-Thierry and St. Mihiel. Liggett had obsessively studied the 1870 Franco-Prussian War, fought in the very same peaks and ravines. The rugged officer knew the path to Germany better than anyone else in Pershing's command. This had already served them well, when Liggett devised the daring rescue of the Lost Battalion. The tall, imposing Pennsylvania native emerged from his mid-October meeting in Pershing's second floor office at Souilly "his face glowing; his eyes sparkling as though he had seen a vision come true," observed war correspondent Frederick Palmer, waiting nearby.[40]

Coincidentally, just a few moments later, an aide ran up the stairs to deliver the message that the new German chancellor had accepted Woodrow Wilson's Fourteen Points as a basis for peace negotiations. It appeared Germany did not want to fight forever. Berthe Hunt heard the news first. It was she who put the call from G2 (Intelligence) through to Pershing's staff. "My, what pandemonium—we phoned the news everywhere," she wrote. "Such excitement" as word spread through the command.[41]

Their jubilation proved premature. Marshal Ferdinand Foch insisted on unconditional surrender. Yet the end was in sight. For the next three weeks, deadly new artillery drives on both sides alternated with rumors of German capitulation. The British liberated Ostend, Belgium, on October 17. The Germans bombed towns near First Army headquarters, including Ligny. The women were told to expect their turn next. Yet Souilly was miraculously spared. Enemy fliers did not know the location of First Army headquarters. Telephone operators had helped to preserve the secret, aware that enemy agents monitored many conversations, trying to deduce where Pershing was holed up.

Hunter Liggett and his doughboys held the enemy's feet to the flame, while Berthe Hunt and other operators fed the precise time of day to artillery units calibrating their shots and connected calls to a nearby airfield, where pilots like Eddie Rickenbacker waited by the telephone. Tens of thousands more died while government officials decided how to end the war. Ironically, the Allies did not wish to push so hard that the German government might fail altogether, leaving no one with whom to make peace.[42]

Word came over the phone lines on October 28 that Austria had surrendered, yet still the Germans fought on. Grace Banker learned that First Army headquarters might soon be moved again, trailing the quickly advancing AEF. Banker and her girls would accompany them. She stayed up until 4:30 A.M.

on October 26 to welcome her replacements, and arrived at the office the next day with eyes so bloodshot she was ashamed to be seen.

The next two days were wild for Banker and Hunt, orienting five new women drawn from Neufchateau, Langres, and Chaumont. To create more living space, the Signal Corps transferred their switchboards across the mud-slicked alley to another wooden barracks. Sawing, hammering, and banging compounded the chaos at the exchange, where every electrical connection was rerouted. More lines were added, and the women took over additional switchboards from the French. Berthe Hunt trained the inexperienced operators, who griped at the uncomfortable accommodations, while Grace Banker fed everyone the new codes they needed from two sheets of paper the engineers pushed at her. Berthe Hunt worked eleven hours, then went back to her quarters to study the revised codes.

Captain Mack appeared a day later at Souilly. He brought Grace a beaded purse from Domrémy, the birthplace of Joan of Arc. "He is certainly thoughtful," Grace wrote in her diary, as if not sure what his gift meant. Earlier that month, when the Australian dropped by tired and hungry from the front late one evening, she fixed him bread and jam from a stash she kept in the exchange. He told her he might have to go to the United States to bring over more troops, for which he would need American papers. "I don't know how that can be," Banker wrote in her diary, missing the possible hint that he hankered after some more permanent connection with her country.[43]

As the Hello Girls' chief liaison with the army, Banker's prominence required her to demonstrate immunity to the sexual implications of female integration. This responsibility, upon which the success of women in the army depended, and the Victorian norm that placed the initiative on men to express romantic intentions, may have handicapped her in interpreting Captain Mack's feelings—or reaching out more assertively to an officer she appeared to adore.

Adele Hoppock had a sharper eye for relationship possibilities than Banker and perhaps more freedom to ruminate on them. Lieutenant Marmion Mills, stationed at Ligny-en-Barrois, called long distance to ask if she would mind bringing him some candy from the commissary in Neufchateau on the eve of her transfer to Souilly. Not thinking anything of helping a buddy, she hastened to the commissary, knowing her car would pass through Ligny. "It dawned on me afterwards," she wrote her mother, "that the reason he wanted the candy was so we would have an excuse to stop in and see him on the way."[44] Lieutenant Mills subsequently "happened by" Souilly more than once to see Hoppie—whom he later married—which explains why he was present the day their barracks caught on fire.

First Army headquarters used German prisoners of war as orderlies. Perhaps as an act of sabotage, one of them overturned a stove in the old wooden offices of Army Intelligence, G2, shortly before noon on October 30, 1918. Adele Hoppock had just ended her shift and was chatting with Lieutenant Mills when the two noticed black smoke rolling out of a nearby building. The fire did not make an immediate impression until Colonel Parker Hitt dashed around the corner. Grace Banker was just leaving the new exchange when Suzanne Prevot yelled out, "Looks like a fire."[45]

The flimsy barracks were tinderboxes, with their oil-paper windows and tar-paper insulation. The sun shined brightly that morning, one of the few on which it did not rain that frigid autumn. The conflagration sent up great funnels of smoke and flame, then leapt to other buildings in the same row. The fire started licking at the edges of the Signal Corps building.

"Get what you can out," Colonel Sosthenes Behn urged Grace Banker. She and others not on the switchboard ran into the barracks, followed by soldiers who rushed from all sides. They grabbed what they could, throwing blouses, pictures, papers, food, and toiletries onto blankets that they quickly bundled. A duo shoved the precious piano out the door. One man ran out with a mattress under one arm, a water pitcher, gas mask, and toothbrush in the other. "Not a minute too soon," Banker wrote. "Like an exploding shell, our barracks burst into flame." Heedless of her belongings, all she could think was, "How lucky the office had just been transferred to the other barracks."[46]

But that structure was in danger, too. Choking smoke poured into the switchboard room, where Berthe Hunt and the others were working. Shouts rang out above the quiet voices coming over the wires. "Of course we sat at the board & operated—there was nothing else we could do," Hunt wrote. "We knew our things were going." But they also knew they could not break contact with embattled frontline troops.[47]

A bucket brigade tried to save the switchboards. Colonel Behn climbed onto the roof over Berthe Hunt's head and emptied pails of water that others handed up. The normally immaculate Behn was streaked with soot. Grace Banker watched from outside in horror. A new drive was scheduled for the next day. How would they communicate with crews positioning their guns at the front and the one million men clawing their way through the forest? "The very thought was too terrible," she realized, hoping against hope the exchange could be saved.[48]

The fire grew hotter, and the smoke thicker. The operators ignored the danger. Colonel Parker Hitt later reported to the War Department, "The girls on duty at the board remained there until it seemed impossible to save the building." Berthe Hunt wrote in her diary that night, "At last they came to us & told us to pull out our connections & escape by the back door." Colonel Hitt ordered the engineers to haul the precious switchboard machinery into a nearby field. For the first time during the entire war, General Headquarters stopped talking with the front. The other operators dashed to an empty lot and watched as soldiers struggled against the fire.[49]

Seven buildings were reduced to ash. Miraculously, the bucket brigade saved the new telephone exchange. Linemen immediately reinstalled the switchboards and repaired the cut lines. An hour later, headquarters was at least partially reconnected with the rest of the army. Berthe Hunt and her colleagues were back at the switchboard.

After work the women wandered the camp to find their belongings, scattered among the ruined buildings. Grace Banker located her toothbrush in a shoe, her prayer book atop a piece of steak in a frying pan. Berthe Hunt found a bag with random items gleaned from her cubicle, including three eggs she had bought the day before to make a birthday cake for another soldier. There was "n'er a crack in them," she noted with amazement.[50]

Homeless, the operators found new quarters in an unfinished building. Grass peeked through hastily placed floorboards. Logistics (G4) invited the women to share their mess. The women appreciated the hospitality and rations, though Hunt later discovered, "Our S.C. officers are as jealous as can be." Artillery invited the women as well, but logistics beat them to it. When G4 asked if they could have the rescued piano, Artillery retorted, "they can't have girls & piano too." Berthe Hunt laughed to herself. "It is funny!" she wrote.[51]

Not everyone found the situation amusing. For the next twenty-four hours, the women fielded complaints about having been irresponsibly incommunicado. An aggravated Grace Banker tied vainly to appease the new chief operator at Chaumont, who would not stop chastising her. "We were forbidden to tell anyone what had happened for if it gets out the Germans may put two and two together and realize that First Army Hdqts. are at Souilly." It was hard for women in the rear to comprehend difficulties at the front, Banker later wrote. "They lived in the midst of Peace. We lived in the midst of War."[52]

For every soldier who entered that arena, it took awhile to adjust. The only complaint registered against the Hello Girls occurred toward the end, when one of the newest operators at Souilly failed to note a temporary change in command. The front moved so quickly in the final days that troops sometimes outdistanced the artillery barrages that were supposed to crush enemy defenses before they arrived. Colonel George C. Marshall was acting chief of staff on November 6 and dispatched two sets of pilots to ascertain the position of troops on the ground. When he learned that his men had run ahead of their assumed position, he reached for the telephone.

"The receipt of this information demanded instant action," Marshall wrote in his memoir. He requested an immediate connection to artillery, to order them to stop firing, "but was told by the young lady at the switchboard that the line was busy." Marshall demanded she yank the plug and put him through. The Hello Girl told him that only the commander in chief or his chief of staff could interrupt a call. When Marshall pointed out that he was acting chief of staff, she "quibbled."[53]

Marshall did not know this officious telephone operator, but her instincts were dull and information old. Marshall ordered her to connect him to Colonel Parker Hitt, which she did. Parker promptly ran to artillery headquarters next door and told them to stop firing. Conscience stricken and upset, the operator called Marshall to apologize and later came to his office to inquire "how much damage her action might have caused." Marshall opined, "This was apparently the first time any of the women centrals at Army Headquarters had been brought into such immediate contact with the direction of the battle."[54]

Yet Marshall was uninformed, too, and in an acting position. Veteran operators of First Army Headquarters had been in contact with the battle from the outset. They understood that each call might be a life saved or lost. Just like men, Hello Girls faced a learning curve, and this operator was green. War couldn't be taught in the abstract. It must be lived.

Five days later, everything was over. America's late start had been fumbling, but the army's might was now substantial and lethal, providing "the margin in manpower and materiel that allowed Foch to wage his war of attrition successfully."[55] Seasoned British and French troops broke through Germany's so-called Hindenburg Line. Kaiser Wilhelm II abdicated. The Allies offered an armistice on the basis of Wilson's Fourteen Points with stringent requirements, such as the right to occupy Germany. Pershing believed they could extract an unconditional surrender but considered an armistice acceptable as long as it prevented Germany from ever again taking up arms.

On the evening prior to the signing, Berthe Hunt worked at her usual station when an Artillery lieutenant came to the switchboard office at 8:00 P.M. to tell her that all fighting would stop in fifteen hours, at 11:00 A.M. "My, the shouting and racket in camp when the boys heard of it," Berthe Hunt wrote in her diary. Then she sat up through the night, a water bottle at her feet for warmth, listening to the final barrages and taking the last calls of the Great War.[56]

The next morning, a captain of the Signal Corps telegraph office found his line was down when he tried to telegraph the news across the front. The men must get the announcement before 11:00 A.M. In panic, the captain ran to the switchboard office. Grace Banker handed him her own headset, and he yelled the message to everyone he reached. The vast scale of the front and rudimentary communications meant word did not get to everyone in time. Some men kept firing, and others were killed moments after the eleventh hour. An army nurse within shell range learned of the armistice only from the silence. She was still receiving wounded men when "there was a sudden silence about 11 o'clock in the morning. We wondered what was different."[57]

Pershing's chief operator felt stunned at the war's abrupt, arbitrary conclusion. "We have lived so long under war conditions that it doesn't seem that it could come so simply," Grace Banker wrote. The French communications officers with whom she and her operators collaborated left that very afternoon amid smiles and tears. One of them, a Sergeant Alexandre, solemnly shook each woman's hand and wished her good luck. Then, pressing his cap to his heart, he said simply, "Au revoir, Mademoiselles," and was gone out the door.[58]

According to British historian John Thompson, America "determined the outcome of the European war."[59] Pershing had proven that the United States would carry through to the end. The Hello Girls had, too. Bell Telephone proclaimed, "For the first time in history a great war has been carried to a successful conclusion with women recognized officially as part of the army."[60] Only the War Department seemed not to care.

11

PEACE WITHOUT THEIR
VICTORY MEDALS

A LINE OF ECSTATIC French civilians danced past Merle
Egan's office in Tours singing *Les Marseillaise*. They
waved at her to join them, but an officer ordered her to remain
at her post, keeping the swamped telephone lines open. Fire-
crackers banged and sparkled. "The French had gone mad,"
according to Merle. "French soldiers on leave were kissing
every woman on sight." At Souilly, Berthe Hunt watched
filthy American troops tramp past First Army Headquarters. "Boys coming back from the front—such joy you never
saw!"[1]

Victory was sweet. Excitement, relief, and optimism were
in the air. But what had been won? Would a new world organization rise from the ashes and make all nations safer? For
the wounded, widowed, and orphaned, the transition to nor-
malcy was hardest. For some it was impossible, such as the
cousin of Louise LeBreton, who lived out his remaining years
at Les Invalides and covered his face with a cap to hide the
remains of his face whenever she visited.[2] Yet daunting complications arose for everyone, including Woodrow Wilson and
the female soldiers of the U.S. Signal Corps.

They, and the suffragists, had to fight on.

On December 2, 1918, the day before he sailed for Europe to negotiate the peace treaty, Wilson gave what turned out to be his last State of the Union address. Over the next year, the fight for the League of Nations consumed the president's attention and ruined his health. Yet before he embarked on his epic quest to establish a new world order based on arbitration rather than aggression, Wilson made a supreme effort to finish old business. Women's suffrage was number one on the list prepared by the president's Irish American secretary, Joseph Patrick Tumulty. The Democrats must pass the Susan B. Anthony Amendment before the newly elected Congress met in the spring or Republicans would get all the credit.[3]

Wilson's farewell meeting with Congress was an awkward affair that presaged difficulties to come. Three weeks earlier, when Wilson read the armistice aloud in the ornate House chambers, Republicans and Democrats cheered. Now rivals sat silently while Wilson explained his reasons for attending the peace conference. No president had traveled outside the Western Hemisphere while in office. More importantly, Wilson was heading to Europe alone—that is, without any of the politicians whose vote he needed on a treaty. Wilson did not invite a single senator to Paris to reengineer America's place in world politics. The president's overweening desire to control events likely cost him his dearest wish. A man who cared less might have accomplished more.[4]

Wilson opened with the one subject on which all could agree: the "audacity, efficiency, and unhesitating courage" of American men who had risked their lives. "No soldiers or sailors ever proved themselves more quickly ready for the test of battle or acquitted themselves with more splendid courage and achievement when put to the test," Wilson proclaimed to general approbation. "Those of us who stayed at home did our duty," he remarked in the code of gallantry, but would always consider themselves "accurs'd" not to have served at Château-Thierry, St. Mihiel, and Meuse-Argonne. "The memory of those days of triumphant battle will go with these fortunate men to their graves." Some had already gone.[5]

"And what shall we say of the women?" Wilson asked, pivoting abruptly. They had shown "instant intelligence" in the emergency and made every challenge easier. "Their contribution to the great result is beyond appraisal," the president asserted, stating a fact with which no legislator could openly disagree, even if he resisted its implications. "The least tribute we can pay them is to make them the equals of men in political rights. . . . These great days of completed achievement would be sadly marred were we to omit that act of justice."[6]

Suffragists kept up the chorus in ensuing weeks. Women had served. Gender must no longer be a qualification for full citizenship. Six days after Wilson's address, Carrie Chapman Catt presided over a mobbed, jubilant gala at Washington's National Theatre to honor war workers. Hundreds had to be turned away. Clothed in the wartime uniform of the YWCA, the wife of Secretary of the Navy Josephus Daniels urged passage of the suffrage amendment so that women could contribute to making the "world safe for democracy." Woodrow's youngest daughter, Eleanor, told listeners that "the history of American women during the months of the great war is the last word, the direct and conclusive proof of their fitness for self-government and for full citizenship in the business of the nation and of the world."[7]

International news stoked the sense of momentum. Revolutionary Russia had granted the vote even before the

Bolsheviks took over. An editorial cartoon in a Columbus, Ohio, newspaper in 1917 showed the star-spangled wife of Uncle Sam pointing to Mrs. Russia's stylish "equal suffrage" hat and complaining, "And to think you let her get it first."[8]

When the war ended, Christabel Pankhurst and thirteen other British suffragists announced their candidacies for Parliament under the new suffrage law. Returning doughboys brought back the news that America lagged. One soldier ran down a gangplank and asked a group of women waiting at the bottom with sandwiches and coffee, "Have you got it yet?" They replied, "Got what?" He answered, "Why, the vote." When they told him no, he retorted, "O hang, you ought to be ashamed. The German women have it."[9]

The Senate took up the question after Wilson left for Europe, where he encountered grateful, adoring throngs from London to Rome. Wilson wrote Joseph Tumulty repeatedly from Paris to inquire, "Is there anything else I can do that might help to bring about the passage of the suffrage amendment?" Members of Alice Paul's National Woman's Party meanwhile pressed relentlessly. On February 9, the day before the next Senate vote, they burned the president in effigy in front of the White House.[10]

The following morning, a throng of supporters watched the Democrat-controlled Senate from the galleries. Once again, South Carolina promised to provide the last swing vote. Another death had brought William Pollock, a man beholden to Wilson, into the Senate. It was there that the newest Democratic legislator gave the keynote address reintroducing the constitutional amendment.

Women had proven themselves fit for citizenship in multiple ways, Pollock said. If nothing else, their decades-long fight for the vote showed political courage. Women had persevered despite "censure and ridicule." The president himself had "appealed to this body to do this simple act of justice to the mothers, the wives, and the daughters of this land just as Britain has done."

Most important, women had "doubly earned" their reward, Senator Pollock said, echoing the now familiar mantra equating military service and citizenship. They had laid aside comfort, pleasure, and safety to help the nation, taking up "man's work in order that men might be free to fight for liberty and civilization." Yet another speaker, Republican Senator William Calder, pointed to nurses who had been "with our Army from the beginning of the war, at times under most difficult circumstances, and some have been under fire repeatedly."[11]

The Senate still did not budge. Suffragists, the president, and supportive congressmen shaved the vote down to one holdout, but the amendment failed a second time in less than five months. Racist southerners provided the bulk of no votes, citing states' rights, yet die-hard opponents from the northeast helped, too. Henry Cabot Lodge quietly knifed the joint resolution along with every senator from Massachusetts, Connecticut, and Pennsylvania. The Sixty-Fifth Congress closed for business without the win Wilson longed to give his party.

Suffragists remained undaunted. More remarkably, the president kept plugging, too, though militants continued to jeer him, and he had already lost the chance to keep the majority of seats in Congress. On March 10, the National Woman's Party rallied thirty-five hundred enthusiastic supporters in New York to cap the cross-country tour of their Democracy Limited—Prison Special train, from which those who had been jailed for civil disobedience gave whistle-stop speeches. They reiterated their arguments at posh Carnegie

Hall, where audience members sang "Les Marseillaise," now an anthem of freedom worldwide. "The United States fought for democracy," one speaker argued, "and who got it? Our enemies. Women fought for democracy and received mockery. The German women have been fully enfranchised and thirty-four of them are seated in the German parliament." Another accused Wilson of insincerity. Had the president really cared, his party would have passed the amendment.[12]

But the president could not merely order the legislature around, and his influence had waned. Even before the Sixty-Sixth Congress convened in May 1919, it looked to become a bloody battleground over the Versailles treaty. After Wilson shared a draft of the proposed covenant of the League of Nations with congressmen in early March, Henry Cabot Lodge organized thirty-nine of forty-nine incoming Republican senators to pledge their opposition. Before the treaty hit the floor, Lodge had more than enough votes to defeat it in Congress.[13]

As Joseph Tumulty observed, Henry Cabot Lodge nursed a "strong rooted dislike," even a bitter hatred, of Woodrow Wilson, and in some ways the men were opposites.[14] Lodge took mostly conservative positions, whereas Wilson hewed to the Progressive line. Lodge was a Boston aristocrat who looked down on southerners, Wilson the son of a Virginia preacher. Lodge was primarily a nationalist, Wilson primarily an internationalist. The senator emphasized that Americans should "not look to foreigners in order to find out what they think," whereas the president believed the United States could learn from others.[15]

Yet they shared traits, too. Both were PhD academics fiercely convinced of their own opinions, and both wanted the United States to play a more active role in the world. Lodge was more rigid by temperament, but a series of strokes in 1918 rendered Wilson more brittle than Lodge. Nonetheless, the New Englander was the worse snob, and his pettiness knew few bounds. As Lodge said of the president's authorship of the Covenant of the League of Nations, "It might get by at Princeton but certainly not at Harvard."[16]

Lodge consistently voted against women's suffrage, though it was not his paramount concern. He declined to speak during the debates, perhaps to avoid alienating progressive Republicans. After the Republicans won their one-vote majority in November, Lodge assiduously cultivated other party members. He ignored previous clashes on domestic questions. Above all, he wanted their votes on foreign policy.[17]

At the time, the Constitution required Congress to adjourn on March 4 in odd-numbered years. The president had no obligation to reopen the legislature until the following December. Yet Lodge stampeded Wilson into an early start by filibustering on appropriations in late February. If the executive branch wanted operating funds, it would have to call the new Congress earlier than statute required. The president took up the challenge. Wilson used the occasion not only to ask for appropriations but also to plead once more for women's suffrage.[18]

The president sent his message from Paris, where he was attempting to modify the treaty in ways that might mitigate Republican objections. On May 20, an aide rose to read Wilson's letter. "Throughout all the world this long delayed extension of the suffrage is looked for; in the United States, longer, I believe, than anywhere else," Wilson wrote, alluding to the birth of suffrage agitation in 1848, when such unconventional ideas required "steadfast courage." The request was succinct, but he had already made it repeatedly.[19]

Since the joint resolution had been defeated in the prior session, both houses had to vote again on the amendment. Opponents embellished all the old arguments. Senator Oscar Underwood of Kentucky called the federal amendment an attempt to "invade the family home and fireside." Representative Benjamin Focht of Pennsylvania quoted a female anti-suffragist who had written him that only a "noisy minority are demanding votes for women." Wives and mothers desired no reward for their sacrifices in the war effort. The esteem of menfolk was tribute enough, Focht implied, appealing to the patriarchal assumption that "normal" women abhorred public recognition. Congressman Frank Clark of Florida, who later scoffed at the wartime contributions of African American veterans, agreed that "the good, pure woman, the queen of the American home," simply did not want the ballot. "The men of America have protected them for a century and a third, and they will do it yet."[20]

Proponents emphasized two arguments specific to the moment.[21] The United States had fallen out of step with other great nations, and female patriots had proven their worth in war. Congressman Edward Little of Kansas City, the ranking Republican on the Women's Suffrage Committee, observed that men had "for centuries treated woman as a slave, dragged her over the pages of history by the hair, and then you pretend to think she is an angel, too good to interfere in the affairs of men." But women were neither angels nor slaves. They needed acknowledgment as citizens, and a framework of law within which to defend their own rights if abridged. "Give her now a fixed, reasonable status, as becomes a rational human being like yourself."[22]

Congressman John Raker of California pointed out that "practically every civilized country in the world has extended the right of suffrage to women." Was it reasonable, he asked, that "we . . . who boast that we stand for giving men the opportunity to express their voice in our Government . . . should be the last?" Former majority leader Champ Clark of Missouri concurred, "We can not afford very well . . . to be behind every other civilized nation in this matter." Republican Congressman James Mann of Illinois, author of the bill, pronounced, "You can no more stop the wave of progression than you can roll back the sea with a broom." As Puritan leader John Winthrop had cautioned three hundred years earlier: "the eyes of all people are upon us."[23]

And there was the question of wartime service. "Everywhere you went during the past two years you saw women in uniform," Congressman John MacCrate of New York reported. "Whether you were home or whether you were abroad . . . you realized that American womanhood had met the last argument that men have given for denying them the suffrage privilege, namely that no one who is not a potential soldier is entitled to the franchise." Uniformed women had earned full citizenship. Adolphus Nelson of Wisconsin agreed: "The time is now ripe." Women had "demonstrated in the awful war just ended, as well as in every crisis of the world's history, her undisputed right to equal suffrage."[24]

The bill zipped through the House in a day. The year before, representatives had passed the amendment by a single vote. This time, the margin was a comfortable forty-two votes beyond the required two-thirds.[25]

The Senate traveled a more twisted path. The debate lasted five days, and larger criticisms bubbled up like gas from an awakening volcano. Southern racists were not the only ones to oppose federal activism.

Republican senator William Borah, a leading liberal, challenged the Susan B. Anthony Amendment on the grounds of
states' rights. Calling himself a supporter of suffrage though
he opposed the bill, he was the only senator to criticize the Jim
Crow measures that had turned the Fifteenth Amendment
into a "dead letter." Since the Nineteenth Amendment would
be "a solemn lie," too, why write it onto the books? No black
woman would gain the vote. The catastrophic Civil War was
within living memory, and a new law might prompt yet another showdown that would injure African Americans more
than whites, or so Borah asserted, assuming that national co-
mity required leaving blacks to their fate. "Sphinxlike, in-
scrutable, and intractable. . . . There is no phase of national
life, no outlook but is colored by the sinister shadow of this
problem." Known as "the Lion of Idaho," William Borah enjoyed roaring at southern racism but had no stomach for using
federal power to enfranchise women of either race. In the
coming months, he would also become the leader of the "irrec-
oncilables," those unalterably against the League of Nations.[26]

New York senator James Wadsworth, another opponent,
agreed. The proposed amendment would enfranchise black
women in name only. States would make the law a "dead
letter." Better not to pass an unenforceable statute than foster
"contempt for the Constitution." These words were music to
Senator Ellison Smith of South Carolina, who warned against
being "stampeded by a few hysterical propagandists." Drawing
on Greek mythology, which blamed females for destroying
paradise, Smith said suffragists would open "Pandora's box of
evils." Any vote for the Nineteenth Amendment reinforced
the Fifteenth, and "there is not a man in America today capable of exercising the functions of citizenship but that rec-

ognizes that amendment . . . jeopardized the civilization that you and I represent."[27]

Senator James Reed of Missouri concurred: "We might just as well look this question in the face." Interracial solidarity between females was potential dynamite. "When the clamors go up from the dark sisters of the South that they are not being permitted to vote, and the sisters of the North who belong to a political party that feels it is losing votes down South gets aroused," Reed fulminated, "I want to say to you, Senators, you are very likely to get some legislation compared with which the force bill [during Reconstruction] will be a gentle and merely persuasive measure." Reed urged legislators to consider the amendment's champion—a president willing to jettison such hallowed traditions as George Washington's policy of no "entangling alliances."[28] Wilson believed women's suffrage would reinforce his global policy. Men like Reed feared he might be right.

Republican Frank Brandegee of Connecticut trumpeted similar criticisms of the president and his foreign sympathies. Brandegee reminded the Senate that the president had previously respected states' rights. "Now, having resided for the last six months in a foreign country [France], he cables to his subservient idolators here how they shall vote on this constitutional amendment and they will 'come to heel' with due humility, I have no doubt." Brandegee became an irreconcilable as well.[29]

Suffrage proponents had less to say. Their arguments had been aired for seventy years. More important, they finally had the votes. When roll was called on June 4, 1919, the Senate passed the amendment 56 to 25. The galleries echoed with applause. The very next item of business was a critique of the Treaty of Versailles.

With Wilson's continuing support, two-thirds of the states ratified the Nineteenth Amendment within a year.[30] Ten states, all southern, turned it down. Nonetheless, votes for female citizens became law on August 26, 1920. By then, twenty-six other nations had approved women's suffrage, and Wilson had suffered the massive stroke that disabled him and helped lead to the defeat of the Versailles treaty. The United States rejected membership in the League of Nations, with fateful consequences when Japan and Germany attacked their neighbors in the 1930s. Despite the president's high hopes for U.S. participation in a new structure of international law, the vote for women was his last great legislative accomplishment.

Carrie Chapman Catt later mused over why the first nation to consider women's suffrage was one of the slowest to adopt it. Much of the resistance arose from sectional and partisan politics within a nation still healing from civil war and in the midst of mass immigration. But underneath was a more profound reason, at least in Catt's view: a kind of "biological foundation of male resistance." The decades had shown "how slowly men as a whole retreated from the 'divine right of men to rule over women' idea, and how slowly women rose to assume their equal right with men to rule over both."[31]

The fight against chattel slavery had introduced an additional impediment, as if greater freedom could be granted to one group only by diminishing another. In 1868, the Fourteenth Amendment introduced the word "male" into the U.S. Constitution. It gave full citizenship to former slaves, and required Congress to reduce the number of seats apportioned to any state that denied the vote to "*male* inhabitants" over the age of twenty-one. In other words, nonvoting adults could no longer be counted as three-fifths of a person for purposes of representation. If a state refused to allow black men to vote, it could not include them in the population numbers that determined seats in the House of Representatives.

Although Congress failed to enforce this provision of the Fourteenth Amendment, the legislation explicitly denied a constitutional protection to women for the first time, making clear that racial justice superseded gender justice. Feminists protested that the Republican Party had just pitted women against former slaves. Reformer Wendell Phillips merely replied, "One idea for a generation, to come up in the order of their importance. First, Negro suffrage, then [alcohol] temperance, then the eight-hour movement [for workers], then woman suffrage." Women were last.[32]

Indeed, major reforms arrived in more or less the order Phillips predicted, and the Nineteenth Amendment eventually rectified the Fourteenth. But it took a long time. Catt observed:

> To get the word male in effect out of the constitution cost the women of the country fifty-two years of pauseless campaign thereafter. During that time they were forced to conduct fifty-six campaigns of referenda to male voters; 480 campaigns to urge Legislators to submit suffrage amendments to voters; 47 campaigns to induce State constitutional conventions to write woman suffrage into State constitutions; 277 campaigns to persuade State party conventions to include woman suffrage planks; 30 campaigns to urge presidential party conventions to adopt woman suffrage planks in party platforms, and 19 campaigns with 19 successive Congresses. . . . It was a continuous, seemingly endless, chain of activity. Young suffragists who helped forge the last links of that chain were not born when

it began. Old suffragists who forged the first links were dead when it ended.[33]

Once the vote was won, some women questioned its value. Voter participation declined overall during the 1920s, as a cynical view of reform swept not only the United States but also the world. The postwar generation retreated from activism, recognizing the lives wasted by prideful nationalism, the futility of reforms like alcohol prohibition, and the chaos unleashed by the collapse of the Russian, German, and Ottoman Empires. Women eager to participate in political parties were shunted into "women's committees" appointed by men.

Party leaders planned it that way. As the Republican state chairman told the *New York Times* on the eve of the 1920 nominating convention, "There will be no trouble worth mentioning from the women who will sit as delegates in the convention, and there will be about 200 women delegates—more than have ever sat in a convention. But they will all be good organization women." Not for another fifty years did women start to make trouble, when they again branched off from male-dominated groups to form their own.[34]

Feminists made a few legislative gains in the early 1920s, such as the Cable Act, which allowed women who married foreigners to keep their citizenship, but after this there were virtually no victories until the 1960s and 1970s on gender-related issues that mattered to postsuffrage organizations like the League of Women Voters. In the South, black women were largely unable to exercise the vote, as prophesied. They enthusiastically joined the electorate whenever possible, but labored under the double burden of racism and sexism.

Scholars call this the legacy of disenfranchisement. Laws change, but they have a half-life.[35] It often takes institutions a generation or longer to catch up. Women and African Americans required several more decades (and a new civil rights movement) to marshal substantial political influence. Yet possession of the constitutional right to cast a ballot, even when it did not have immediate meaning, proved potent. The Fourteenth, Fifteenth, and Nineteenth Amendments were rungs on the political ladder, giving women and blacks a place from which to climb. African American suffragists showed cognizance of this by actively supporting the Nineteenth Amendment, though well aware of southerners' plan to foil the new law.[36] In a legalistic society, words had power. Thus, getting them on paper mattered.

There was a parallel in foreign relations. The U.S. Senate rejected what Wilson called "a machinery of good will and friendship" to prevent war. Opponents of the League of Nations argued that Europeans need not worry if America failed to sign the treaty, since, as one senator opined, the United States could be "trusted again to come to the rescue" if trouble surfaced. But such reassurances proved hollow. Relieved they had not subscribed to collective security, Americans walled themselves off as long as possible from the violence rampant in Asia and Europe in the 1930s, until violence shot from the sky over Pearl Harbor.[37]

Woodrow Wilson warned that the Senate might "break the heart of the world" if it failed to pledge its commitment. The League of Nations dissolved in tears beyond measure.

While Congress was fighting its last battles over women's suffrage, Signal Corps operators remained under army discipline in France. Close to two million doughboys and officers had to be repatriated following the November armistice. The process required expert coordination. The army replaced females with enlisted men wherever unusual expertise was no longer required, but retained them for the most critical assignments. A team was soon ordered to Coblenz, Germany, to run telephones for the U.S. Army of Occupation. Woodrow Wilson's peace commission needed service before and during the Versailles conference. All 223 female soldiers of the U.S. Signal Corps were still in France for Christmas in 1918.

The event was a high point. AT&T donated cash to Signal Corps houses to purchase gifts, and the AEF gave each woman a memento booklet filled with letters and photographs from officers expressing appreciation for their "skillful performance of vitally important work [that] . . . contributed in no small measure, to the success that has crowned our arms." Adorned with glossy 8 × 10 photographs of Pershing and Russel and bound with lavender cord, the books expressed the men's surprise, awe, and gratitude.

Brigadier General Edgar Russel's letter, dated the day after the armistice, acknowledged that skepticism had attended the women's arrival. "It pleases me a great deal," he wrote, "to say that by your ability, efficiency, devotion to duty and the irreproachable and business-like conduct of your affairs, personal and official, you have not only justified the action taken in assembling you, but have set a new standard of excellence that could hardly be improved upon." Women had established a "brilliant reputation" that he expected them to sustain until the army no longer needed them.[38]

Dozens of other officers, including twenty brigadier generals, added their thanks for the women's "grit," "unfailing courtesy," ready acceptance of "hardship and privations," "cheery responses to gruff and unreasonable demands," "unremitting toil," "indispensable service," "unsurpassed" performance, and "wonderful success." Tank commanders, quartermaster units, and ordnance officers outdid one another in heartfelt letters, most excerpted, some reprinted in full, filled with Christmas wishes and expressions of admiration for the "valiant telephone operators." Brigadier General Fox Connor, chief of staff for G3 (Operations), wrote that not only had the women been indispensable, but their conduct "constantly encouraged us to emulate your patience and devotion to duty." Brigadier General George Van Horn Mosely from G4 (Logistics) called them "better disciplined than the Army itself."[39]

Some praise echoed gender stereotypes of females as helpmates, but the overt message was that uniformed women had performed beyond the call of duty. "They are entitled to the highest commendation," base commander Colonel John Sewell wrote. L. H. Bash, an adjutant general with Services of Supply, seconded his assessment: "They have rendered a service in this war that the country will be proud to recognize."[40]

Colonel Parker Hitt filed his official report two days after the armistice, when events were still fresh. As chief signal officer for the First Army, he told superiors "a large part of the success of the communications of this Army is due to . . . a competent staff of women operators." The women outperformed almost every man in the job. "They had a most uncanny way of finding routes to strange places and it seemed impossible for them to give up a call after it had once been filed [requested]." The American Expeditionary Forces were "indebted to these women who gave their labor so cheerfully for its success."[41]

Over the next several months, senior officers selected thirty of the 223 Signal Corps operators for special commendations, most signed by General Pershing. Merle Egan, who ran the exchange for the American Peace Commission at Versailles, Louise LeBreton, Berthe Hunt, Adele Hoppock, and many others earned the ornate citations. Louise LeBreton's was typical in its specificity, recognizing "exceptionally meritorious and conspicuous services at Headquarters, Advance Section, S.O.S., Neufchateau, France."[42]

Grace Banker earned the highest honor of all. On May 10, 1919, the Decorations Section of the U.S. Army notified the chief operator that the AEF "Commander-in-Chief, in the name of the President, has awarded you the Distinguished Service Medal." Woodrow Wilson had created the award only the year before, along with the Distinguished Service Cross, for bravery in combat. General Hunter Liggett, commander of the First Army, pinned the gold and enamel medal to Banker's uniform at a full military review in Coblenz, Germany, where she was stationed with her beloved brother, Corporal Eugene Banker, who had survived gassing. The citation praised the chief operator's service from Chaumont to the Battle of Meuse-Argonne. "By untiring devotion to her exacting duties under trying conditions," it read, "she did much to assure the success of the telephone service during the operations of the First Army against the St. Mihiel salient and the operations to the north of Verdun."[43]

American newspapers spread the word. City dailies and rural journals from the *Philadelphia Inquirer* to the *Pueblo Chieftain, Augusta Chronicle, Fort Wayne News-Sentinel, Kansas City Star,* and *Duluth News-Tribune* reported that America's first female soldier had received the army's highest honor. "First Woman to Enlist in U.S. Army Gets D.S.M. for Heroism during War," the *Denver Post* told readers. "The ceremony took place at Coblenz, in the presence of a great gathering of this brave little woman's fellow Americans in the army of Uncle Sam." The *Post* noted that Banker's recently widowed mother learned the news from cable dispatches, and that the Barnard College graduate had been recognized for "splendid work, sometimes under fire, in organizing the telephone service of the first American telephone unit in France."[44]

When interviewed, Banker asked reporters to "say just as little as possible about me." Credit was due the "dandy, faithful and ever ready" women of the Signal Corps, who were all heroes. Banker went so far as to express impatience at all the attention. Her work, she offered, did not compare with the hardships and achievements of the men under fire. This might be interpreted as the attitude of a self-deprecating female if it did not echo the phraseology of most officers when discussing the sacrifices of combat troops.[45]

Black Jack Pershing thought Grace Banker merited special praise, however. Of 16,084 eligible officers in the Signal Corps, she was one of only eighteen to win the DSM. At the end of 1919, the chief signal officer in Washington published their names in order of rank, from Major General George Squier down to Colonel Parker Hitt. Grace Banker was listed last, her rank given as "Miss," the only title for female soldier in the army's lexicon. In its final report, the Signal Corps notified Congress that, "a total of 323 decorations—81 officers, 243 enlisted men and 1 woman telephone operator—have at this date been awarded in the Signal Corps by the United States, French, English, Belgian Governments."[46]

Signal Corps operators meanwhile continued under orders. Some applied to go home but were told that they must stay until "relieved from further duty." Unlike YWCA or Red Cross volunteers at liberty to depart, uniformed operators were required to remain on the job. When the Signal Corps suggested relieving women from duty, Lieutenant General Robert Bullard, commander of the Second Army and the hero of Cantigny, protested vigorously in mid-February 1919.

Bilingual female operators were essential to the "complicated and urgent" business of the Second Army, charged with preserving order, guarding lines of communication, and salvaging equipment in the chaotic, postconflict environment. Bullard told the Signal Corps, "I consider female operators necessary for efficient handling of service with which they are familiar."[47]

Louise LeBreton, one of those most frequently pressed into translating for the Second Army, requested a discharge that same month due to a family emergency. Although her supervisor approved the request "as soon as the services of Miss LeBreton can be spared to the Signal Corps," it was another two months before Louise was allowed to leave. The War Department finally sent a letter clarifying that "her discharge has been amended to take effect April 29, 1919." Merle Egan requested separation in early March 1919 and was also denied. Two months later, without warning, the army notified Egan that her discharge had been granted. She must report to Brest in five days.[48]

Berthe Hunt was desperate to see her husband—"can scarcely wait for time to pass"—but it wasn't until two weeks after the armistice that she finally received leave to visit him. While waiting, she went with other operators during breaks to explore the front lines they had faced for months, scarred by thousands of Swiss-cheese shell holes now iced over, quarry-like dugouts, abandoned trenches, burnt hills that looked like pictures of primitive "cliff dwellers," and "graves, graves everywhere." When they had a free morning to drive south, green meadows and working farms reappeared after a few hours. Suddenly, miraculously, the women found themselves amid "peaceful little villages everywhere & such good roads," Hunt wrote. "It seemed so queer," she told her diary, "it is so peaceful here." Everywhere, they met heartbroken mothers.[49]

Finally, the day for which she had waited almost a year arrived. Hunt and her husband rendezvoused at her uncle's home in Paris. They must have flown into each other's arms. "Oh, it was good to be . . . with Rube!" she wrote that night. They took a train south and had an ecstatic week in Nice until she was summarily ordered to Paris along with Grace Banker. "Lucky Rube can go where he wishes," Hunt told her diary; "the Navy isn't as particular as [the] Army."[50]

In Paris, Hunt received a promotion to supervisor, for which she retook the army oath one more time. Yet with danger at an end, work became monotonous. Her Signal Corps family was "breaking up," with Parker Hitt and other officers ordered to new posts. "Want to get out," she wrote from the rainy capital. On Christmas Eve, Hunt scribbled, "Oh, this work I hate—just plain hello work, finding rooms for stranded officers & arranging dates in the evening." Hunt had signed up for soldiering. Like many veterans, she found peacetime jarring. She also wanted a transfer to Brest, where her husband's ship was docked, but the army needed her in Paris. The Signal Corps brought almost half of its operators, eighty-four women, to the capital to run communications.[51]

Hunt penned the last entry in her wartime diary on December 31: "Do hope the New Year will see me and Rube together more than 5 weeks." The army did not discharge her for another four months, on April 2, 1919.[52]

Grace Banker found "the petty squabbles of civilian life" insufferable. She was given charge of the exchange in Woodrow Wilson's Paris residence, a position of high honor, but "work at the President's house was not particularly exciting." Her wartime nerves must have still been jangling. Banker took the first opportunity to join the Army of Occupation in Germany. "We missed the First Army with its code of loyalty and hard work," she said about herself and others with whom she had served. Five months later, in September 1919, Banker finally boarded a troopship for New York.[53]

The last operators did not leave until 1920. Cordelia Dupuis of North Dakota, who slid down a hotel bannister to a Paris air-raid shelter in March 1918, was finally discharged on January 20, 1920, with three AEF stripes on her sleeve and an award for meritorious service.[54]

One operator who never returned was the Michigan recruit who swore the army would have to build a bridge before she would cross the Atlantic again. Corah Bartlett was thirty when she sailed with the sixth unit. In her army ID photo, the uniformed operator wears military-issue steel-rimmed glasses and a serious expression—though a photograph before the war reveals a smiling young woman in Victorian-era dress. Bartlett was still serving when she died of typhoid fever at a U.S. military hospital on June 23, 1919, one of tens of thousands of soldiers who perished from disease exposure. She was buried in her Signal Corps uniform and dog tags at AEF Cemetery Number 33, Indre-et-Loire.[55]

An honor guard led the parade to the graveyard in Tours. A single officer walked in front of the coffin covered with long-stemmed roses. Soldiers marched behind, accompanying Bartlett to the open grave beside a long row of white wooden crosses that marked the resting places of other Americans. She remained there until her brother asked the army to bring her home three years later, when her remains were repatriated along with twenty-one other "deceased soldiers." Bartlett was then given a second military funeral in her Michigan hometown.[56]

Personnel records like Corah Bartlett's were sprinkled with terms that hinted at the ambiguity of the women's status. Some documents called them soldiers, others civilians. Most described them as "discharged" or "relieved from duty," others as "terminated" or "deceased." Adele Hoppock was discharged under orders identical to those issued to other soldiers in August 1919 with one distinction. Her base commander specified that because of the twenty-one-year-old's front-line service, she was "authorized to wear the Victory Medal and one Defensive Sector Clasp, for the following operation: Meuse-Argonne, September 26th-November 11th." Four million men were entitled to the Victory Medal after the war. Far fewer could claim the Meuse-Argonne clasp.[57]

Hoppock was one of only a handful of Hello Girls to receive such decorations in 1919. Corah Bartlett and Inez Crittenden were the only operators accorded a military funeral at the time. Women had voter registration cards by 1920, but none had the army discharge papers fundamental to public recognition as veterans.

12

SOLDIERING FORWARD IN THE TWENTIETH CENTURY

EMALE VETERANS typically discovered they were not soldiers in the eyes of their government—and thus the world—when they applied for something, whether treatment at an army hospital, membership in a veterans' organizations, tax benefits offered by local municipalities, or the bonus granted to every soldier, sailor, marine, and nurse. Surprise characterized their letters.

Polite Christine Bickford, Merle Egan's roommate in Tours, petitioned General George Squier when the American Legion threatened to expel her. She knew he was busy, but could he please send her papers? Insouciant Hope Kervin of San Francisco dashed off a request for a copy of her discharge because "The tax people seem determined to have it." Impoverished Louise Chaix pleaded for her veteran's bonus as "I have three children to raise." Irene Gifford, too ill to write for herself, was the subject of correspondence among military officials unsure if she qualified for government hospitalization. Tuberculosis contracted in France had settled in Gifford's sixth dorsal spine, fourth rib, and left lung. "Disability originated in the line of duty," one surgeon noted. "Disability is total and permanent."[1]

Whatever they applied for, the Hello Girls received the same baffling answer. They had been contract employees even though they had not signed contracts. They were not entitled to a single postwar benefit, not even the name veteran. Discharge papers would not be forthcoming no matter how many times their release was described as a discharge. They had not been soldiers no matter how many officers had told them, "You're in the army now."

That was almost the exact phrase Merle Egan heard when she applied for transport back to Helena, Montana, upon disembarkation in New York on June 2, 1919.[2] A quartermaster said he could route her to Denver. Egan insisted, "I don't want to go to Denver. I want to go to Helena."

He retorted, "Young lady, you are still in the army."

As former head of the U.S. Peace Commission switchboard in Paris, she answered, "Yes, sir! And I have been in the army long enough to know that I must be discharged at the point of enlistment if desired." When the officer muttered something like, "damn insolence" under his breath, Egan added, "Moreover, sir, we were treated with respect by our officers overseas."

The man flushed, then corrected himself. "I beg your pardon. I will have the transportation order changed."[3] Perhaps that was when Merle Egan became the champion of the Signal Corps operators.

The Westerner was an interesting bundle of contradictions. She was an authoritative, no-nonsense woman with a strong sense of purpose. Child of a unionist, she had a feisty streak and did not hesitate to stand up for what was right. When Egan and Grace Banker both roomed briefly at the Signal Corps house in Paris, they volunteered to represent either side of the political divide at animated household meetings: "Miss

Banker heading the Conservative [Party] and Miss Egan the Bolshevists," an amused YWCA minute-taker noted.[4]

But on personal matters, Merle Egan followed a path defined by male initiative. When she joined the army, the man courting her gave her a peck on the cheek, no "real embrace," and said nothing about marriage. While Egan was in France, several officers wooed her ardently. Each excited her more than Hal Anderson, and she fell deeply in love with a fellow Montanan—"a young Adonis" in his army uniform, she thought—who appeared to share her feelings. But Egan remained uncertain of her obligation to Hal Anderson, who wrote infrequently. Then, in February 1919, his letters suddenly became affectionate and insistent.

As Merle later learned, he had heard from mutual acquaintances of her romance with the officer from Montana. Hal Anderson admitted that he was determined to "spike" the competition. He pleaded, "all the troops are coming home, why can't you?" Anderson's girlfriend had become famous. Montana newspapers reported in January 1919 that one of the state's "cleverest" girls had been appointed chief of the Versailles exchange. Hal Anderson urged Merle to quit Wilson's peace commission and return to Helena to marry him. Merle was unsure of her feelings. The other officer had not declared his love, but she had also not made plain that her prior relationship was over and she was available. Feeling obliged to follow through since Hal Anderson had finally offered marriage, she applied for a discharge, granted two months later.[5]

Once Merle disembarked in New York, she received yet another letter from her hometown suitor, this time indicating that the wedding might have to wait for financial reasons. Upset and confused, Merle almost called off the engagement. She longed to escape to Paris—or follow Grace Banker to

Germany. But Merle had applied for a discharge to get married, and it made her feel oddly compelled. "I was so imbued with army regulations," she later recalled, "that I had to marry someone!" Coping with the disorienting transition from wartime to peacetime, Merle Egan numbly followed the path of least resistance. Three days after her train pulled into Helena, when the war was still a fresh wound, Merle followed through on her pledge to the Assistant Traffic Supervisor she described as a boring, stocky, stateside army reserve officer who snored. Hal Anderson wanted to wed immediately after all. He was an acceptable, decent man, though not the one who made Merle giddy with joy. Despite her ambivalence, they spent their lives together.[6]

Yet Hal Anderson gave Merle a renewed purpose. Proud of his wife's war record, Anderson suggested she write the War Department in July 1919 for her Victory Medal. Depressed, Merle declined to do so. He wrote for her. The official reply that she had not been "in the army" failed to rouse Merle. Then she learned that the navy had granted full benefits to female yeomen, including hospitalization for the disabled. The idea that the War Department would depreciate her sisters in the Signal Corps changed everything. Merle Egan Anderson became a leader once more. She would not make the wrong choice on a fundamental question a second time, especially when it came to defending or serving others.[7]

Merle's ensuing fight exposed not only the War Department's aversion to change, but also the remarkable dedication of those men who knew the operators best. Officers' wartime experiences of responsible women performing vital work converted at least some Signal Corps brass, much as it had Woodrow Wilson. Men's loyal advocacy proved critical during the next sixty years. Even more important was women's initiative. Although some declined to fight their government's harsh judgment, others opposed it with the grit they had shown since the day they first raised their right hands.

The signatures on the armistice were hardly dry before General Edgar Russel began planning "the discharge of female telephone operators." At AT&T in New York City, R. F. Estabrook clarified that they should be given military transport back to their point of origin, and that "following their discharge, operators are entitled to wear the uniform without insignia for a reasonable period of time."[8]

The same week in early February 1919 that the first few trickled home, General George Squier asked the army's top legal counsel to remind Congress about the operators when they doled out bonuses. Everyone should be recognized, including those disabled before they could sail, as well as the 111 uniformed operators waiting at the dock in Hoboken when the war ended. Squier pointed out they had "the same status as Army nurses."[9]

Two weeks later, on February 24, 1919, Congress passed legislation granting a "gratuity" of $60 to "all persons serving in the military or naval forces." State legislatures soon voted gratuities, too, most in the vicinity of $100. In 1924, Congress tacked on a second bonus with the Adjusted Compensation Act. The reward was substantial. Veterans would receive $1 for every day served in the United States between 1917 and 1919, and $1.25 for each spent on a foreign shore. Veterans like Grace Banker, enlisted 20 months, would earn roughly $750, close to a year's salary for most working women.[10] The federal bonus was an interest-bearing note payable in 1945, though the Great Depression later spurred some veterans to demand immediate redemption in the infamous Bonus March of 1932.

Adjutant General Joseph Trany unexpectedly spurned Squier's request to honor all Signal Corps operators, including those disabled during training. Secretary of War Baker "does not consider that female telephone employees have the same status as army nurses," the lawyer informed Squier, so "no special legislation will be requested making this act applicable to these employees."[11]

George Squier objected immediately. He reminded the lawyer that the War Department had specifically promised operators a status akin to nurses, who also served "by appointment and not by enlistment." Even if considered civilians, operators were guaranteed the privileges "now prescribed or which may hereafter be prescribed by Army Regulations and General Orders for Army nurses." Two weeks later, Trany sent a terse rejection that depersonalized the women altogether: "The class of civilian employees referred to in this communication does not come within the provisions of the Act."[12]

The War Department told its lawyers what to say. Newton Baker wished to avoid financial responsibility for an even larger group of women who had also risked their lives. If operators were honored, the department feared, Red Cross canteen workers, ordnance stenographers, quartermaster clerks, and Salvation Army nurses who went to France might be next. The War Department must hold the line. If Congress acknowledged the operators, "then all female employees . . . [might] be included."[13]

Virtually all operators were still stationed in France when bonuses were first dispensed to men who had returned, with the result that women at home noticed first. Those discharged from Hoboken were especially disappointed. Some had trained in distant cities nine months for no purpose. Many had quit good jobs to assist their country. All had purchased expensive uniforms. In May 1919, the seventh unit signed and sent a petition to the War Department. A friendly Chicago businessman (the father of an operator) took the matter up with Senator Medill McCormick, a partner in the *Chicago Tribune* who helped pass the Nineteenth Amendment a month later. ATT wrote the War Department, too.[14]

In their petition, operators requested bonuses like those received by other soldiers who had remained stateside, and reimbursement for the $300 uniforms that were "a total loss for us." Perhaps in its "multitudinous duties," they charitably speculated, the War Department had simply "lost sight of the work of the Women's Telephone Unit and their rights." Nonetheless, "our pride is severely injured in being ignored."[15]

Lower echelon officers spent the rest of 1919 working through the tangle of regulations that even General Squier had been unable to tame. Their vocabulary reflected the challenge in making women's service sound like civilian employment. One Signal Corps colonel acknowledged to Senator McCormick that the uniforms were "unsuitable for wear upon their discharge," but he justified the expense as if they were sporty togs from which women had gotten fair use. Uniforms "were worn for some months while in the employ of the government before the operators were discharged," Colonel C. Saltzman reassured the perturbed Illinois senator, having employed the military term "discharge" twice.[16]

Even the Civil Service Commission was confused and wrote Newton Baker in October 1919, "What is the difference in status between Army nurses and members of the telephone unit?" The commission had never issued contracts to the women, so presumably it had no disability obligations either.[17]

The Signal Corps was forced to justify a policy it detested. Nurses were subject to a fuller version of the Articles of War than operators, the Corps replied vaguely, giving no examples. Operators were like "retainers to the camp." Camp retainers had always and everywhere been subject to court-martial. "Female telephone operators were not subject to military discipline other than in the general sense" that anyone was when attached to the army.[18] In other words, the skillful, indispensable operators were somewhere on the spectrum of camp followers.

Other officers refused to give up. Major Robert B. Owens, who first suggested female operators to Pershing and Russel in the summer of 1917, wrote the army to ask "if anything can be done to further recognize the invaluable work done by American women operators with the A.E.F." The Hello Girls deserved whatever benefits pertained to "officers, men and nurses." Owens added, "I believe General Pershing is of like opinion." Signal Corps Colonel Aubrey Lippincott replied that he had been advised there was "absolutely no possibility of special legislation." Yet the corps would not forget, Lippincott promised. A later Congress might grant "the recognition these girls so justly deserve."[19]

General George Squier tried to get the adjutant general to issue Victory Medals, also without success. In 1922, he made a third run at veterans' benefits, this time following the army quartermaster's lead. The SOS had sent approximately two hundred women to France as surgeons' note takers, stenographers, laboratory technicians, dietitians, and medical clerks during the last months of the war. They worked alongside enlisted men performing identical duties. Some took pay cuts of 50 percent compared with their civilian earnings.[20]

In 1922, the quartermaster asked the new Republican secretary of war, John Weeks, to support legislation granting
them and Signal Corps women the same benefits as yeomanettes. Squier appended his grievance to the quartermaster's.
Female operators, he wrote, "rendered conspicuous service
during the war under trying conditions and exposed to the
hazards of submarine and aerial warfare, and are deserving
of recognition of their splendid and unselfish service."[21]

Once again, the quest went nowhere. An internal report
from 1926 indicates why. The War Department considered
women "not only unnecessary but troublesome to an Army."
It also wished to avoid financial exposure. Between large contingents at domestic cantonments, and tiny units deployed
abroad, the army had used over 100,000 women during the
war "in unorganized and uncoordinated groups hastily and
inefficiently recruited . . . with no military status or recognition." Compared with the Department of the Navy, which had
enlisted, trained, and discharged 13,000 women, the army
was a mess.[22]

The War Department was downsizing, and had no desire
to pay benefits to anyone beyond the four million men already
eligible. Yet even if all 100,000 non-uniformed women had
been added, the veterans' roll would have increased only
2 percent. Money alone did not motivate the War Department. Stubborn pride, bureaucratic arrogance, and the belief
that women simply did not merit recompense blinded senior
staff officers to faceless female veterans.

Men who knew those faces persisted. When a former operator informed Major Roy Coles—who had approved Inez
Crittenden's transfer and commended Merle Egan in Paris—
that some states questioned whether operators deserved

bonuses, he wrote an impassioned letter in 1921 to each state board. Coles had supervised uniformed women every day. "I am in the very best position to know of the sacrifices and devotion to duty shown by these heroic young ladies who placed themselves subject to the strictest kind of military discipline . . . and shared the hardships and privations which come to an army in a theatre of operation." Coles acknowledged it had not been practical to "commission or enlist these young ladies because they were females," but Pershing, Russel, and everyone who knew them would resent "as keenly as I do any intimation that these ladies were other than a part of the American Expeditionary Forces in the fullest sense of the words and of the phrase."[23]

Others chimed in. Captain Ernest Wessen, who went to work for Congress in the early 1920s, approached the chairman of the Committee on Military Affairs. The congressman told him, "Ernest, there is no doubt in the world but that your telephone operators were combatants, and should have been bona-fide members of the military establishment, but if we provide for this tiny group at this time we shall be forced to reopen the cases of thousands of other applicants."[24] Wessen was told to come back in a few years.

Major Stephen Walmsley, another Signal Corps officer who had supervised women on a daily basis, went to bat with the American Legion. Although the women had technically been civilians, Walmsley told the legion, they served wherever sent and functioned as "an integral part of the Headquarters 1st Army and Headquarters 2nd Army." Some withstood bombardment. All risked submarine attack. In 1934, Franklin Delano Roosevelt's chief signal officer urged Congress again to pass appropriate legislation. As he knew from personal experience, female operators deserved "the highest commendation," Major General Irving Carr wrote.[25]

The impasse settled into a pattern that persisted another fifty years. At least twenty-four times between 1927 and 1977, representatives in Congress introduced bills to acknowledge female veterans. The War Department blocked each attempt. No bill made it past the Armed Services Committee. When Adele Hoppock Mills requested a copy of the War Department's correspondence with Congress in 1935, the former operator received a snippy reply.

Major General E. T. Conley of the adjutant general's office could not reveal internal correspondence, but he did not mind telling the decorated Meuse-Argonne veteran that the army "has always been opposed" to acknowledging those who entered the service by irregular means. Women like her had been "very fairly compensated." If operators were recognized, Conley wrote, it would open the floodgates to other civilians in "semi-military" capacities. Why should the pushy Hello Girls "be singled out for preferential treatment" denied others equally deserving? As the War Department argued when similar legislation pended fifteen years later, even the Knights of Columbus would have to be included![26]

The erasure of the operators' wartime contribution was nearly complete by 1936, when Major General Conley dismissed Maude McMullen's request for a bonus as well. Without a peek at personnel files that contained hundreds of sworn statements, the general blithely told McMullen she had never been under oath. Operators were not "required to take the oath subscribed to by members of the army or navy." They had "agreed to accept from the United States the pay offered," and in view of that, General Conley declared, "the

War Department is without authority to establish your military status as you now desire."[27]

Whether motivated by malice or ignorance, the effect of this untruth was the same: it denied the Hello Girls' patriotism, sacrifice, and service record.

Male supporters had tried. Women had to carry the offensive forward. When the American Legion began expelling female veterans, they formed their own organization in 1921. The Women's Overseas Service League (WOSL) accepted all who had served abroad, refusing to discriminate against Red Cross, YWCA, and other volunteers. The blurring of private and public work had disadvantaged females disproportionately.

As Merle Egan Anderson later explained to army secretary Cyrus Vance when she served as WOSL's national chair for legislation, "In 1918 the term 'civilian army' was a common expression. The whole Army was 'civilian.'" To operators told they were the first female soldiers, "the fact that some of our papers used the word 'civilian' had no meaning for us."[28] But when the war ended, females were held to the most stringent definition of "civilian," unlike males drawn into the reserve, such as John Carty. WOSL applied to the U.S. Congress for a federal charter similar to one granted the brand-new American Legion, only to be turned down in the U.S. Senate. The women started a magazine and named it *Carry On.*[29]

Grace Banker, Berthe Hunt, Merle Egan, Olive Shaw, and the LeBreton sisters all appeared on the 1931 membership roll, but Signal Corps operators also maintained a separate identity beyond WOSL. Private letters reflected their enduring bond. At Christmas in 1930, Merle Egan Anderson sent each "Friend and Buddy" a list of married names and new addresses to help them keep in touch. "We have no record of our Y.W.C.A. Hostesses and we feel that they should be included," she wrote from her new home in Seattle. "Will someone please forward such a list if they have it?"[30] In the 1940s, Grace Banker, Suzanne Prevot, Merle Egan, Louise LeBreton, the Hoppock sisters, and others exchanged greetings at holidays and birthdays.[31]

Some were more active than others in pressing for government recognition. Grace Banker, married to a man who did not serve in France and the busy mother of four children, shied away from further activism. Yet the wartime experience remained important to her, and she carefully preserved her gas mask, trench helmet, tin mess kit, and letters in the green leather-bound Signal Corps trunk that she took to Paris in 1918. When prompted, she wrote a short article for AT&T on the service of the telephone operators in which she praised her girls but refused to glorify war. She called it "the world's worst plague," and noted, "one glimpse into an Evacuation Hospital would make a pacifist out of anyone."[32] Louise LeBreton and Adele Hoppock had fewer or no children and were married to former doughboys who understood what the country owed veterans. It may explain their willingness to fight on. Merle Egan was encouraged by her husband, though the marked contrast between their wartime accomplishments created some awkwardness.

The army itself was slow to learn lessons. When World War II broke out, the service again found itself scrambling. It immediately recruited women to help in multiple capacities, some even as pilots. Like telephone operators, pilots were uniformed, sworn in, and placed under military discipline, but not officially enlisted. The war lasted four years instead of one, however. There was time to organize, and institutional memory about what had happened in 1918 survived. The army came under immediate pressure to act more honorably than it had twenty years earlier.

Although War Department bureaucrats opposed militarizing females, Congresswoman Edith Nourse Rogers (for whom former Signal Corps operator Olive Shaw worked as an assistant) introduced legislation to prevent another "tragedy" that left disabled female veterans without hospitalization benefits. General George C. Marshall, who knew the army needed women's skills, supported Congresswoman Rogers. WOSL organized a campaign to get women "*in* the Army." In 1942, Congress created an official Women's Army Corps and commissioned women as officers. Anyone enlisting after that date earned full recognition. African Americans were allowed to serve, though segregated. They were officered by other black women, two of whom rose to the rank of major.[33]

After the war ended, the Hello Girls of 1918 lobbied Congress more persistently. Merle Egan Anderson sought the advice of Ernest Wessen, retired from public service and enjoying minor notoriety as a specialist in rare books.

"You cannot imagine how happy I am to hear from a member of one of the old Telephone Units," Wessen replied. He warned her not to make common cause with women whose civilian status had been uncontested. "*You folks were actual combatants,* and so recognized by Pershing when he decorated one of your number for bravery under fire." Major Wessen stressed the point repeatedly in subsequent letters. Under the same laws that made uniformed operators combatants, "the Red Cross is an international body of recognized non-combatants. They haven't a shadow of a claim." Wessen advised against muddling Signal Corps claims with those of private organizations or women paid for clerical work. Unfortunately, this was easier said than done, since the army itself refused to recognize such distinctions.[34]

Merle Egan Anderson had written Wessen in the context of a renewed campaign. In January 1947, Ohio Congressman Charles Elston introduced a bill to acknowledge women who had served "in uniforms provided, approved, or prescribed, by the Army of the United States." Secretary of War Robert Patterson urged the Armed Services Committee to reject the bill. Too much time had passed, Patterson said. It would be hard to determine who had worn a uniform. The cost would be prohibitive. It would "establish a precedent" for other civilians.[35] None of these objections had much basis in fact, but they convinced war-weary Congress.

Robert Taft of Ohio, a strong contender for the Republican presidential nomination, introduced a Senate bill to recognize Signal Corps veterans in 1950. Congresswoman Edith Nourse Rogers of Massachusetts followed it up with another in the House. The Mississippi American Legion backed the proposed legislation "unalterably, unequivocally, and without reservation." The Washington and Ohio state branches expressed support, too. A Massachusetts representative read an editorial from *Stars and Stripes* into the *Congressional Record* arguing that there was no comparison between private volunteers and "devoted and patriotic" women in uniform. "It is late, but never too late to make amends."[36] Former operator Olive Shaw supplied Congresswoman Rogers with arguments to use in her testimony.[37] The bills failed again in committee, and the Korean War distracted Congress from further efforts.

New bills were introduced in the next decade, when the reformist John F. Kennedy came into office. Opposition gathered steam, too. The Veterans Administration and the Bureau of the Budget both strongly advised Congress against recognition as "discriminatory" towards other "civilians."[38]

Signal Corps veterans refused to give up. Christie Bickford Newbert of Maine—Merle Egan's former roommate—convinced her representative to file a new bill. Then she petitioned Senators Edward Kennedy, Mike Mansfield, Everett Dirksen, and Hubert Humphrey. All replied with standardized letters bearing original signatures that pledged "sympathetic attention" and a willingness "to expedite consideration" should the bill come to the floor. Vice president Lyndon Johnson sent a reply that was "as smooth as a kitty's ear," Newbert told Merle. Johnson promised the operators that although he no longer participated in Senate debates, he would keep their concerns in mind. Newbert's representative forwarded her letter to the young president, himself a disabled veteran of World War Two.[39]

The bill failed again. Merle initiated a friendly correspondence with Pulitzer Prize–winning civil rights journalist Hodding Carter, hoping he would help. She petitioned Secretary of Defense Robert McNamara and General William Westmoreland. Harold Say, the editor in chief of *Stars and Stripes*, became a friend and staunch advocate. The newspaperman suggested others to contact. Outside the circle of women, Merle sometimes felt Harold Say was "the only person who really listens or cares." When her representative from Washington State submitted a new bill in 1967, Merle again reached out to all Signal Corps buddies, urging them to write their congressmen or women. "It is probably our last chance." She also challenged yet another Defense Department spokesman who rejected the evidence: "Is there no way to penetrate that stone wall of prejudice?"[40]

Some men cheered on the women; others did not. The brother of Oleda Joure of Michigan wrote Congress, as did Colonel Marmion Mills, husband of Adele Hoppock Mills. But Louise LeBreton Maxwell's husband still considered the women he served alongside as having merely relieved men. They had made no special contribution. Berthe Hunt's husband, Rube, liked to annoy his wife by calling the Women's Overseas Service League, to which she was fervently devoted, "the Shell Shock Sisters." After Grace Banker's untimely death from cancer in 1960, her friends recalled the uncommunicative Columbia College engineer she married in 1923 as a "dour Scotchman."[41]

Legislative efforts failed in the reformist 1960s despite dozens upon dozens of supportive letters. Merle Anderson began to tire. Her husband Hal had died the decade before. Turning seventy-nine, she investigated a retirement home. She told friends she did not have her old "pep," though she continued to serve as finance chairperson for the Women's Overseas Service League and co-organized its forty-fifth annual national convention.[42]

Entering her eighties, Merle wrote the commander of the men's World War I veterans' organization, "I realize my time, also, is short but I feel our small group deserves, at least, a place in Army history." Discouraged, she penned Louise LeBreton, with whom she exchanged frequent letters, "Isn't it ironic that Alger Hiss [tried for perjury] has a place in [the Columbia] Encyclopedia yet Grace Banker, first woman [soldier] to receive D.S.M. is not listed?"[43]

Yet Merle could not quite manage to give up. "My reason may sound queer," she wrote another former operator, "but, I love my country, consequently, I want that country to be worth loving."[44] The disaffection of the turbulent sixties and seventies, and the weakening of popular faith in government, troubled her. She had followed the government's rules: thrown her diary overboard, eschewed earrings, married when she said she would. Washington must honor its own promises to sustain people's trust and affection. She told Michelle Christides, the daughter of Oleda Joure, "When I was asked recently what I would like to see most before I die, I said: 'I would like to see the young people say, 'I love my country,' as they did in 1918.'"[45]

For Merle, the message of old age was not to slow down but to hurry up. She got a renewed burst of energy. Her correspondence suddenly quickened. Slim files grew thick. She contacted Chet Huntley of CBS, NBC's *Today*, and *Reader's Digest*. She convinced *Yankee* magazine to reprint an excerpt from Grace Banker's short memoir, "I Was a Hello Girl." She gave interviews to local newspapers and spoke to schoolchildren. "Our only real chance is publicity," she wrote LeBreton, who found a veteran to donate her uniform to the Signal Corps Museum at Fort Monmouth for a display on World War I.[46]

In the 1970s, Merle Anderson sent a dozen petitions for each one she had written in previous decades, probing for a way to crack the army's case. Polite, cookie-cutter refusals made her madder yet. "I realize that you were trying to be kind in your letter," she wrote Oliver Meadows, staff director of the Committee on Veterans Affairs, when he told her in 1972 that "every group feels its service was unusual" but that veterans' benefits were only for those "who actually served in the Armed Forces during time of war."[47]

Merle's curt reply reveals her frustration. Every petition since the 1930s had contained the same evidence that never penetrated men's thick heads. "Why does Congress ignore the fact that we were told we were 'in the Army,'" she asked, and thus "our Government is guilty of misrepresentation?" Even more galling, "Navy women who served in jobs far less dangerous . . . received those benefits," she fulminated. Numerous male officers had testified explicitly to the service of Signal Corps women in the AEF. "I had so hoped, Mr. Meadows, that you would examine the true facts."[48]

Anderson's letter was dated August 26, the fifty-second anniversary of women's suffrage. Merle Egan Anderson began to wonder if "women's lib" could help. Feminism had revived the decade before with the founding of the National Organization for Women (NOW) in 1966, spurred by the government's decision not to enforce the prohibition against gender discrimination in the 1964 Civil Rights Act.

In order to highlight the absurdity of outlawing discrimination on the basis of race and make legislation harder to pass, Virginia representative Howard Smith had added "sex" to the list. With more females in the population than males, he joked, they already suffered disadvantages—in finding mates. Once again, southerners hoped to sink women and blacks by tying them together. This time, both swam. President Lyndon Johnson backed the provision on women. The act passed.[49]

Yet some still considered women's rights a joke. The male director of the Equal Employment Opportunity Commission, Herman Edelsberg, mocked the provision as a "fluke" that was "conceived out of wedlock," using sexual humor to put down women's rights. Two months after it was founded, NOW began bringing cases to Edelsberg, insisting he enforce the law. They began with flight attendants. As historian Nancy McLean notes, "NOW demanded that the airlines cease treating the female worker 'as a sex object' to be fired when she was no longer judged pleasing to males—when she was no longer nubile, got married, or gained weight." At the time, Pacific Southwest Airlines required attendants to wear pink mini-skirts and fluorescent orange hot pants. National Airlines ran television ads featuring baby-voiced stewardesses who cooed, "I'm Diane . . . Fly Me," or "I'm Marissa . . . Fly Me."[50]

The term "women's liberation" emerged soon thereafter, echoing a phrase used worldwide, from the Palestinian Liberation Organization to the U.S.-based Black Liberation Army. The media immediately shortened it to "women's lib." Feminists wryly noted that no one dared patronize Third World revolutionaries or male African Americans as "libbers," as some dubbed female activists.

Merle Egan Anderson was perplexed by modern feminism. When she saw female marines on television wearing civilian dress and dangly earrings, she was shocked at the younger generation—and jealous. "Had we so appeared in public we would have been Court Martialed. Yet we are denied the veteran status . . . which these Marine women will receive." She wrote Harold Say the month after the debut of *Ms.* magazine: "Perhaps 'Women's Lib' could help us, but personally I would rather have the assistance of the men who served with us and who know the importance of our service."[51]

Anderson's ideas began to change when a new strategy unexpectedly surfaced. Forget courteous requests. This was the heyday of legal activists like Thurgood Marshall, Ralph Nader, William Kunstler, Ruth Bader Ginsburg, and Sarah Weddington. Perhaps the Hello Girls should simply sue.

Charlotte Gyss Terry had drilled atop AT&T with Grace Banker and sailed on the *Celtic*. She remained in Paris when the chief operator left for Chaumont, and recalled into her eighties "when Big Bertha landed every 20 minutes and there were 3 or 4 alerts every night." She still had her Christmas memento book from Chaumont, where Wilson spent the holiday with American troops (including Signal Corps operators), and it bothered her that the army spurned women afterward. Charlotte Gyss Terry did not meet Merle Egan in Paris but wrote to introduce herself in 1972.

She wondered if Mrs. Anderson had read about the odd case of the World War I reserve officers the army sent to Russia in 1917 to keep the railways open before the Bolsheviks quit the Allied cause. After the army discharged them in 1920, the reserve officers were told they had been civilians. Like the Hello Girls, they petitioned Congress. A sympathetic Senate attempted to overrule the army eight times in following decades, but the House of Representatives allowed the matter to reach the floor only once. The men finally sued. In March 1971, a federal court ruled in their favor.[52]

Charlotte Gyss Terry and her husband tried to penetrate the federal law library in West Palm Beach, Florida, to examine the case, but only attorneys could enter. They turned to Merle Anderson. Terry enclosed a newspaper clipping. All she wanted, she said, was "to be allowed an American flag at my funeral services."[53]

Merle wrote back immediately, surprised that Terry had heard of her in the opposite corner of the country. "I am not *too* optimistic—have been disappointed so often—but we *might* just win yet."[54] Merle contacted Harold Say. Did he know about the Siberian railroad engineers? Should she hire a lawyer?

The old newspaperman believed the political process still held hope.[55] Anderson had no money for an attorney, in any

case. At Say's suggestion, she petitioned Oregon congresswoman Edith Green, a longtime supporter, to point out the
example of the railway engineers to the legislature. Anderson
wrote, "I realize that time is short but I am a stubborn Irish
gal (maiden name Egan) who would like to complete this job
before I 'check in.'"[56]

Once firm and precise, her handwriting had become shaky.
Merle joked that she and her typewriter were growing old
together. In April 1974, her only child, Robert Anderson,
fatally crashed his car in the Philippines. A veteran of Korea,
where he earned the Purple Heart, the forty-two-year-old
marine officer was stationed in Manila. Congresswoman
Green, who had introduced four unsuccessful bills on behalf
of the Hello Girls, wrote the grieving mother, "I can well
imagine what a terrible blow this was to you and am most
sorry." Harold Say sent a loving note. "Please know I feel very
deeply for you—a gracious lady whom I have never met but
feel I know exceptionally well."[57] Louise LeBreton and "Suzanne," presumably Prevot, made donations to a Seattle hospital in Bob's memory.

Merle wrote Louise. "I am finding it hard to adjust to the
fact that he is gone. It was so sudden and unexpected." She
dreaded Sundays, the day her husband and son once "pam-
pered" her. At age eighty-five, she told her old friend, "I have
'run out of steam.'" She was almost out of money for stamps
and envelopes as well. After five decades of cordial regards,
Merle and Louise now signed their letters "love."[58]

Yet Merle could still not give up. She sometimes wondered
why she had lived so long, when others she loved were
gone. Like Alvin York, she found herself alone on the front
line, challenged to keep going until she met either victory or
death. Each day brought new possibilities. When President

Gerald Ford proclaimed August 26, 1974, Women's Equality Day, in honor of the anniversary of women's suffrage, Anderson wrote the two newest female members of congress, Republican Marjorie Holt and Democrat Lindy Boggs, widow of the influential House majority leader Hale Boggs, who had recently succeeded her deceased husband. (Boggs later lobbied for women veterans.) "We insist we are victims of 'male chauvinism,'" Anderson told the congresswomen, using the new feminist phrase for the first time.[59]

Early the next year, she wrote the National Organization for Women. Patricia Leeper, co-coordinator of a NOW task force on women in the military, responded promptly. Leeper was the proud daughter of a naval officer. She had applied for women's officer school when she graduated twelfth grade, but was forced to decide between military service and marriage, as married women were not allowed to serve. She chose her husband, a young naval officer with whom she eloped. But Leeper became a feminist when she realized that his identity need not circumscribe hers. Leeper told Anderson she would investigate the matter: "Things may be ripe with the new, more liberal Congress." Pat Leeper began visiting congressional offices about the Hello Girls.[60]

Soon she adopted their cause. She coached Merle Anderson about whom to write on the Armed Services Committee and put her in touch with reporters who could "make the Army look like foolish meanies." She instructed Merle to request operators' personnel records from the National Archives in St. Louis to prove they contained oaths, not civilian contracts. When Republican Senator Barry Goldwater announced he would champion women pilots who had flown for the army at the start of World War II and later been denied benefits, Leeper suggested Merle write him. She did. When Goldwater sent word to Leeper that he would ask the Veterans' Affairs Committee to review the case, the NOW organizer wrote Anderson, "We are quite surprised as Goldwater dislikes 'liberated' women—maybe your case will help us heal that breach."[61]

Pat Leeper's generosity made Merle feel awkward. After she read congressional testimony that her new friend had written about women's future in the army, Merle felt she owed the feminist organizer a confession. Anderson was not sure about the wisdom of integrating West Point and Annapolis, or disbanding the separate Women's Army Corps, given that "nature has given men and women such different bodies." She admitted, "I like being a woman and having men treat me 'a bit special.'" Merle hoped her "audacity" would not affect their relationship. "Am afraid, Pat, the above proves that I wouldn't be a good member of N.O.W."[62]

Leeper swept aside Anderson's caveats. "You are a remarkably open-minded woman. Never apologize for being 'cocky'—we wouldn't be anywhere without women who said their piece and stopped worrying about what other people think." She dismissed Anderson's expressions of gratitude. Fighting injustice "satisfies my Joan of Arc complex," Leeper said, using a metaphor with special meaning to a former Hello Girl. "So you see . . . it is really you who are doing me a favor." A month later, Leeper wrote even more affectionately. "When we get down to the wire . . . I wish you could fly out here. I have a guest room and bath." Pat Leeper began signing her letters with "love" and a women's symbol embellished with a smiley face.[63]

That summer, the NOW leader wrote an editorial for the *Washington Times* encouraging women who thought feminists "hysterical" to soften their judgment of "those of us who may be rocking the boat: You now take the right to vote for granted, yet the very women who gave you that vote were called the vilest of names." When Merle Egan Anderson read the article, she wrote the newspaper's editor, "We need more women like Patricia . . . and more organizations like NOW to encourage her."[64]

A couple weeks later, Merle met a second youthful champion after she gave an interview to the *Seattle Times*, titled "1918 Operator Still Seeks Justice after 50 Years." Mark Hough, an attorney with Seattle's oldest law firm, read the article. He contacted the veteran, then turning eighty-seven, to ask if she needed an attorney. As Hough later recalled, "You meet some people in life who you bond with."[65]

His investigation convinced Hough that the women had a strong legal case and simply needed an advocate to organize and present their evidence properly. He had come of age during the early civil rights movement, and considered racism and sexism equally repellant. The Signal Corps operators required affidavits, personnel records, AEF reports, and internal army documents from 1917 to 1919. Merle wrote Louise LeBreton, "Mark Hough wants to take our case to court." It seemed a reasonable alternative to seeing bills "buried in the Congressional Cemetery" year after year.[66] Merle also put him in touch with Pat Leeper, and continued her own campaign with the media. The American Civil Liberties Union lent its name to the cause.

Mark Hough volunteered his time for the next four years. Using Merle's contacts, he solicited affidavits and other evidence from former operators. The long-retired women mailed mementos they had saved for six decades: wrinkled Army travel orders, telegrams from France, and faded party invitations from other units in Tours and LeHavre. One sent penny-sized binoculars from Paris revealing naked ladies, another an engraved teaspoon she had purloined from a confiscated German cruise ship repurposed as a troop transport. Louise LeBreton mailed her 1918 court-martial notice. Olive Shaw shipped her complete uniform. Hough contacted the attorney for the veterans of the Russian campaign, whose chief plaintiff had been told by an army lawyer, "We'll keep fighting this until you're all dead, so you'll never get what you want." Hough wrote Goldwater, who promised that if the Signal Corps had a case as strong as the Women Airforce Service Pilots (WASPs), he would add them to legislation he had convinced eighteen other senators to endorse.[67]

A former presidential nominee, Goldwater possessed substantial clout. In the 1960s, he spearheaded the rightward thrust that brought conservative Republicans like California governor Ronald Reagan to prominence. Goldwater had also been a pilot in World War II. He considered it outrageous to deny recognition to female aviators, thirty-eight of whom had been killed in this "dangerous and grueling work," simply because of gender. Not that the senator was a feminist. As Goldwater testified, he did not want to "overdo this aspect, but it is a fact that these girls did bear extra burdens simply because they were women." The government owed them recognition not because of their sex but "because they were part of the military."[68]

Merle Anderson felt restored. Younger. "Thank you Mark *for being you*," she wrote the determined attorney. "I feel that you are 'my second family.'"[69]

In May 1977, Merle Egan Anderson received the notice for which she had waited half her life: an invitation to appear before the Senate Committee on Veterans Affairs. The eighty-seven-year-old could not go, but Pat Leeper told the assembly that despite "infirmities," which kept the veterans at home, "they asked me to tell you they are here in spirit."[70] Mark Hough submitted evidence along with his written argument. Everything went into the congressional record, including sworn statements by Merle Egan Anderson, Louise LeBreton Maxwell, Olive Shaw, Oleda Joure Christides, Gertrude Hoppock (sister of the deceased Adele and Eleanor), and four other survivors. They had their say at last.

Hough reminded the Senate that action matters more than intention in law. "There is no question from a legal standpoint that the Army, by treating the Signal Corps women as if they were members of the military, actually made them members of the military, regardless of whatever intention was evidenced in the memoranda of November, 1917." Every officer from Pershing down had acted as if the women were soldiers. Most officers assumed they actually were.

The Russian Railway Service Corps won their case at a cost of $50,000 in lawyers' fees, Hough noted. The army appealed, but the federal circuit court let the decision stand. What had been the court's leading reason? "Members of the Corps wore regulation Army officer' uniforms and insignia. . . . The court noted that the wearing of such uniforms and insignia by nonmilitary personnel is prohibited by law."[71] In short, Olive Shaw's uniform was the crowning evidence.

Mark Hough did not threaten legal action, but the implication was clear. The elderly women could initiate litigation with ample prospect of winning—or Congress could do the right thing. Hough told the women he would take their case as far as necessary, at no cost beyond filing fees.

Once the hearings were over, Hough, Leeper, Anderson, former operators, and supportive onlookers spent the next six months peppering Congress with letters of support. The president of the Women's Overseas Service League wrote the recently inaugurated Chief Executive of the United States: "Many of us voted for you," she told Jimmy Carter. "All of these women are in their late 80s and many are in Nursing Homes. Time is running out. We ask your help . . . so they can at least die knowing their battle was won." She wrote members that Mark Hough was prepared to give "his time (free) to sue for OUR gals." What she needed from "EVERY UNIT of the League is your MORAL SUPPORT."[72]

Some Hello Girls were just too heartsick at the long campaign to trudge one foot further. Ruth Boucher Pels replied that she was not going to do another thing. So few Signal Corps women were left. What did it matter? But the eighty-two-year-old was furious at Black Jack Pershing, whom she thought should be hauled "out of his grave to explain" why he had not done something. "*Oh! How I can't understand Pershing?!?!*", Pels wrote Merle Egan, adding her best wishes and love.[73]

The army and the Veterans Administration still opposed the measure, employing the arguments of preceding decades. Ironically, the comptroller general admitted openly for the first time that there had been no legal restraint against commissioning women as officers in 1918. But since the War Department had not done so then, the women must accept their civilian status now.[74]

The American Legion expressed opposition, too. As its leader explained when Merle Egan Anderson sent a sharp letter of criticism following his negative testimony to Congress, this was not a matter of money but principle. There were too few Hello Girls to cost much. But "to call them veterans would endanger the definition of veterans," Robert Lynch wrote. All sorts of civilians could claim they had advanced various war efforts. Support for veterans' programs would evaporate. Besides, if the army made a mistake, "then it is up to the Army to say so. The Army should be able to figure out who was in the Army and who was not."[75]

But political winds blew differently in the mid-1970s. The women's rights movement enjoyed renewed credibility. Congress passed the bill in November 1977. President Jimmy Carter signed it into law.

"Dear Ladies," Mark Hough wrote, "We have won round one!" It would take two years for the army to process their discharges, but they would eventually receive them. Mark Hough helped thirty-one survivors apply.[76]

Tootsie Fresnel lived to see the day. Grace Banker did not. Cordelia Dupuis made it to the finish. Berthe Hunt did not. Merle Egan triumphed, but Adele Hoppock and three hundred other Hello Girls were gone. Yet they were no longer forgotten.

"Erect and forthright but sometimes trembling with emotion," the *Seattle Post-Intelligencer* reported, Merle Egan Anderson addressed the audience at Fort Lawton when she and two other veterans accepted their discharge papers on August 28, 1979. She wore no earrings, though she, Marjorie McKillop, and Alma Hawkins pinned large white corsages to their bosoms. Their faces were wreathed in smiles. The ninety-one-year-old Anderson declared on their behalf, "The Army has finally admitted we are legitimate." Then she recounted the tale of patriotic service she had told so many times, embellished now by the battle for women's rights. "If I do get a Victory Medal," Merle Egan Anderson snapped, "it should be for fighting the Army all these years!"[77]

Olive Shaw's niece wrote from Littleton, Massachusetts, to thank Mark Hough on behalf of her aunt, who had suffered three heart attacks and could no longer hold a pen easily. "She is so excited about it all, to have it finally gone through, after all these years." The niece wondered if the "faithful" attorney still had her aunt's uniform. "She asks about it." Olive had brushed away tears when they shipped the blue coat, white blouse, and tan parade gloves, "but said if it would help the cause, then that is what she wanted to do." Now she hoped it might go to a museum.[78]

Like Olive Shaw, Mark Hough had trouble letting go. The uniform symbolized the spirited soldiers he had come to love. He considered writing a book about them, and for the next thirty-six years, the World War One uniform hung in Hough's closet while he raised three children, argued law cases in Seattle, and, after the terrorist attacks of 9/11, accepted appointments in the war-torn countries of Iraq and Kosovo to train young attorneys and strengthen judicial systems.

At almost any time, Hough could have put Olive Shaw's uniform in the mail along with the correspondence and memorabilia that Merle Egan Anderson bequeathed to him when she died in 1984 at age 96. The armistice memento book, petitions to Congress, veterans' letters, and faded clippings about death, mud, Jack Pershing, and the Hello Girls could all have been boxed up, delivered to the post office, and set on a plane.

But whenever the Seattle attorney looked at the uniform Olive Shaw loved so fiercely and loaned so reluctantly, he felt he couldn't trust the precious evidence to the mail. Shaw had risked her life for these stripes. Then the Army pretended they meant nothing. Then it turned out they meant everything.

In September 2015, when Mark Hough returned from eight years of international service, he finally hung the skirt, blouse, cap, gloves, and regulation coat with service chevrons and Signal Corps insignia in a moth-resistant suit bag and drove them to Kansas City along with three cardboard cartons. The eighteen-hundred-mile trip to the National World War One Museum gave the seventy-year-old attorney time to say goodbye. There, at last, he committed the record of America's first women's soldiers to the guardians of history.

EPILOGUE

A CENTURY AFTER the Hello Girls and two million doughboys sailed to France, it is hard to picture a world in which telephones look like candlesticks, callers wait for operators to trill bells, and laws deny full citizenship to females and blacks. It seems archaic. But the political and technological trends that began in the late eighteenth century continue to evolve. In 2008, the United States elected its first Chief Executive of African and European descent. In 2016, the Democratic Party nominated its first female presidential candidate, Hillary Clinton. That same year, consumers could ask their wristwatches to make telephone calls.

These advances were not made without purposeful effort. Every woman and African American who served during World War I added momentum to the forward movement. British WAACs, American yeomanettes, army nurses, Russian combat volunteers, Salvation Army lassies, Tuskegee Airmen, and many others, mostly unknown, all played a part. The Hello Girls made an especially important contribution, particularly the seven bilingual women who ensured that General John Pershing never lost the telephone connection to his army. Serving in their navy blue skirts, "upon the very skirts and edges of the battle," they embodied the claim of the 1848 Seneca Falls Declaration of Sentiments: "We hold these truths to be self-evident: that all men and women are created equal."

The experience of the Hello Girls reveals general characteristics of twentieth century reform. Change followed from face-to-face encounters between people otherwise tempted to stereotype one another. The battle for justice advanced through personal relationships. Men who experienced women fighting to pass the Nineteenth Amendment, liberate France, or win World War II were among the first to accept that gender should not define citizenship in a democracy. They helped women reach what became a mutual goal.

The fight to oust the kaiser's army from France unfolded similarly. The United States did not win the war or lose the peace alone. Sustained collaboration among culturally diverse individuals set the parameters of success. Signal Corps operators had to woo their French counterparts over the telephone wires; John Pershing had to earn the respect of Ferdinand Foch. Americans did not dictate to foreign partners, nor could they. Flexibility, cooperation, and adherence to sworn promises led to mutual victory at Meuse-Argonne.

The most successful activists realized that reforms had the best chance when they attracted broad support. Curiously, one way to achieve this was to narrow focus. Justice was difficult to win on multiple issues simultaneously. Carrie Chapman Catt hewed to a single demand that attracted allegiance from people who wanted the ballot for diverse purposes. Suffragists could not challenge racism if they wanted the vote in 1918 any more than advocates of the vote for blacks could readily help women in 1870. The Nineteenth Amendment passed with the support of a southern president and a handful of segregationists. Successful lobbyists were pragmatists, not purists.

As a consequence, the Hello Girls did not bring YWCA secretaries or ordnance and quartermaster clerks along with
them in their fight. They could not tackle the larger challenge
of public–private role confusion and thereby help others who
also deserved better treatment. Although the growth of government in the twentieth century gradually diminished reliance on voluntarism, the Signal Corps veterans of World War
I had to focus on those who wore the brass insignia of the
United States. To win their precious piece of paper they
courted Republicans and Democrats equally. Like Carrie
Chapman Catt, Merle Egan never associated righteousness
with a particular political party.

But while many reformers kept to a narrow focus, their efforts nonetheless had broad effects. The pursuit of justice in
one arena (or one nation) tended to speed its attainment in
others. For example, laws that gave the vote to former slaves,
male immigrants, and lower class Tommies made the disenfranchisement of women seem all the more outrageous. The
comparison made suffragists' arguments pointier in ways that
now appear anti-immigrant and racist, especially in the
United States, since some asserted that educated white
women were better qualified than newcomers and former
slaves—and at least as deserving. Yet reformist movements
mostly assisted one another, and the larger trend of the twentieth century was toward greater democracy and more inclu-
sive definitions of citizenship worldwide. When Oleda Joure
received her discharge papers in 1979, Brigadier General Arthur Wolfe, an African American, delivered them.[1] General
"Black Jack" Pershing would have been astonished—and likely
pleased.

Lastly, reform took perseverance. Progress could be reversed, so it required defending. Prejudice and hubris were

humanity's long shadow. In 2016, four decades after Goldwater's legislation, the U.S. Army denied burial rights at Arlington National Cemetery to the very same pilots on whose behalf the senator had fought.[2] The prized resting place was running out of space, but that was not the reason. The army still failed to take the women seriously. The WASPs, the secretary of the army declared, had never really been military after all. Journalist Cokie Roberts, daughter of Congresswoman Lindy Boggs, registered her protest.[3]

Representative Martha McSally of Arizona, the first female combat pilot, sponsored legislation to overturn the decision. More than 175,000 Americans added their names to a petition. In May 2016, President Barack Obama signed the law that granted WASPs full burial rights. Pilot Elaine Harmon, dead at age ninety-five, was laid to rest.[4]

Later that fall, in the 2016 presidential race, a shocking boast by the Republican nominee that he could grab women's genitalia with impunity came to light. Michelle Obama said, "It reminds us of stories we heard from our mothers and grandmothers about how . . . even though they worked so hard, jumped over every hurdle to prove themselves, it was never enough."[5] Pundits praised the First Lady's speech. Public protest by females had been normalized over the course of a century. But the American people nonetheless elected Donald Trump. Open disrespect for women remained excusable "locker room talk"—and a man who had never held office was elected over a former senator and secretary of state.

Reform and resistance work in tandem to moderate historical change. "Send the word," George M. Cohan sang. The fight for justice continues.

NOTES

ABBREVIATIONS

AHC   Army Heritage Center, Carlisle Barracks, Pennsylvania, World War One Veterans Surveys, Signal Corps, Women Telephone Operators

ATT-NJ   ATT Archives, New Jersey

ATT-SA   ATT Archives and History Center, San Antonio

CR   *Congressional Record of the United States*

Herron   Private Papers of Matt Herron, San Rafael, California

MEA   Merle Egan Anderson Papers, Montana State Archives, Helena, Montana

NARA   National Archives and Records Administration, College Park, Maryland

NPRC   National Personnel Records Center, NARA, Civilian Record Group, St. Louis, Missouri

NWWIM   National World War One Memorial and Archives, Accession 2015.121 (Papers of Mark Hough), Kansas City, Kansas

NYT   *New York Times*

UTSA   University of Texas, San Antonio, Special Collections, Women's Overseas Service League, MS-22

YMCA   YMCA Papers, Sophia Smith Collection, Smith College, Northampton, Massachusetts

PROLOGUE

1. "Miss Merle Egan Chief Operator for Peace Conference Returns to Helena," *Helena Daily Independent*, 15 June 1919, 2.

2. "Hello Girls Here in Real Army Duds: Signal Corps Colors Adorn Hats of New Bilingual Wire Experts," *Stars and Stripes*, 29 March 1918, 1.

3. *CR*, 65th Cong., 2d sess. (GPO, 1918), 10928–10929.

CHAPTER ONE · AMERICA'S LAST CITIZENS

1. Circular No. 27, General Headquarters, American Expeditionary Forces, Office of the Chief Signal Officer, 4 March 1918, Herron.

2. "They're in the Army Now," *American Heritage*, February–March 1980, 111.

3. Lettie Gavin, "Her 60-Year Siege Defeats the Army: Women Vets' Papers Arrive," *Seattle Post-Intelligencer*, 5 August 1979, A7.

4. Wilson, quoted in Kendrick Clements, *The Presidency of Woodrow Wilson* (University Press of Kansas, 1992), 44. See also Daniel Rodgers, *Atlantic Crossings: Social Politics in a Progressive Age* (Harvard University Press, 1998), 56.

5. Woodrow Wilson to Ellen Louise Axson, 31 October 1884, *The Papers of Woodrow Wilson*, ed. Arthur S. Link, vol. 3, *1884–1885* (Princeton University Press, 1967), 389.

6. Wilson, quoted in Frank P. Stockbridge, "How Woodrow Wilson Won His Nomination," *Current History* 20 (July 1924): 567. On Wilson at Princeton, see Victoria Bissell Brown, "Did Wilson's Gender Politics Matter?" in *Reconsidering Woodrow Wilson: Progressivism, Internationalism, War, and Peace*, ed. John Milton Cooper Jr. (Woodrow Wilson International Center for Scholars, 2008), 131. See also Christine A. Lunardini and Thomas J. Knock, "Woodrow Wilson and Women's Suffrage: A New Look," *Political Science Quarterly* 95, no. 4 (Winter 1980–1981): 655–671.

7. Horace Bushnell, *Women's Suffrage; or, The Reform Against Nature* (Charles Scribner, 1869), 30.

8. Millicent G. Fitzgerald, *Women's Suffrage: A Short History of a Great Movement* (T. C. & E. C. Jack, 1912), 5.

9. Mott quoted in Nancy Wolloch, *Women and the American Experience* (Knopf, 1984), 194.

10. Carrie Chapman Catt, *Woman Suffrage and Politics* (University of Washington Press, 1923, 1969), 27; David Lloyd George in Martin Pugh, *The March of the Women: A Revisionist Analysis of the Campaign for Women's Suffrage, 1866–1914* (Oxford University Press, 2000), 189.

11. Lawson quoted in Audrey Oldfield, *Australian Women and the Vote* (Cambridge University Press, 1994), 194.

12. Mineke Bosch, "Between Entertainment and Nationalist Politics: The Uses of Folklore in the Spectacle of the International Woman Suffrage Alliance," *Women's Studies International Forum* 32 (2009): 6, 10.

13. Margot Badran, *Feminists, Islam, and the Nation: Gender and the Making of Modern Egypt* (Princeton University Press, 1995), 75, 92.

14. Mabel Gardiner quoted in Charlotte Gray, *Reluctant Genius: Alexander Graham Bell and the Passion for Invention* (Arcade, 2011), 151–152; Sara Hunter Graham, "The Suffrage Renaissance: A New Image for a New Century," in *One Woman, One Vote: Rediscovering the Woman Suffrage Movement*, ed. Majorie Spruill Wheeler (New Sage, 1995), 164.

15. Grover Cleveland, *Ladies' Home Journal*, May 1905, reprinted in Sherna Gluck, *From Parlor to Prison: Five American Suffragists Talk about Their Lives* (Monthly Review, 1985), 23.

16. New York State Association Opposed to Woman Suffrage, *Thirteenth Annual Report*, December 1908, 10–11, Library of Congress, http://memory.loc.gov/cgi-bin/query/r?ammem/rbcmiller:@field(DOCID+rbcmilscrp5011203).

17. Quoted in Lady Emily Maud Simon, *Women's Suffrage: Some Sociological Arguments for Opposing the Movement* (Cornish Brothers, 1907), 4.

18. Theodore Roosevelt, *The Strenuous Life: Essays and Addresses* (Dover, 2009), 5; Gail Bederman, *Manliness and Civilization: A*

Cultural History of Gender and Race in the United States, 1880–1917 (University of Chicago Press, 1995).

19. Sir Almroth Wright, *The Unexpurgated Case against Woman Suffrage* (Paul Hoeber, 1913), 81–82, http://www.gutenberg.org/files/5183/5183-h/5183-h.htm.

20. Linda Kerber, *No Constitutional Right to Be Ladies* (Hill and Wang, 1998), 236.

21. Ellen Carol DuBois, "The Next Generation: Harriot Stanton Blatch and Grassroots Politics," in *Votes for Women: The Struggle for Suffrage Revisited,* ed. Jean Baker (Oxford University Press, 2002), 170.

22. Blatch quoted in Graham, "Suffrage Renaissance," 165; Alexander Keyssar, *The Right to Vote: The Contested History of Democracy in the United States* (Basic Books, 2000), 159.

23. Carl Degler, *At Odds: Women and the Family in America from the Revolution to the Present* (Oxford University Press, 1980), 331; Keyssar, *Right to Vote,* 54.

24. Degler, *At Odds,* 288, 330, 332–333; Peter Bardaglio, "'An Outrage upon Nature': Incest and the Law in the Nineteenth Century South," in *In Joy and in Sorrow: Women, Family, and Marriage in the Victorian South, 1830–1900,* ed. Carol Bleser (Oxford University Press, 1991), 48–50.

25. Giele quoted in Degler, *At Odds,* 348; Seth Koven and Sonya Michel, eds., *Mothers of a New World: Maternalist Politics and the Origins of Welfare States* (Routledge, 1993), 5; Cokie Roberts, *Capital Dames: The Civil War and the Women of Washington, 1848–1868* (Harper, 2015), 3.

26. Degler, *At Odds,* 347; Jacqueline Van Voris, *Carrie Chapman Catt: A Public Life* (Feminist Press at City University of New York, 1996), 39; Graham, "Suffrage Renaissance," 159.

27. Van Voris, *Carrie Chapman Catt,* 144–145; Keyssar, *Right to Vote,* 171.

28. Graham, "Suffrage Renaissance," 162, 176.

29. Wolloch, *Women and the American Experience,* 246; *Historical Statistics of the United States: Colonial Times to 1970* (GPO, 1975), 1:139–140.

30. Quote from Keyssar, *Right to Vote*, 165; Stephen Norwood, *Labor's Flaming Youth: Telephone Operators and Worker Militancy* (University of Illinois Press, 1990), 143; Lisa Tickner, *The Spectacle of Women: Imagery of the Suffrage Campaign, 1907–1914* (University of Chicago Press, 1988), 57.

31. Catt quoted in Van Voris, *Carrie Chapman Catt*, 72.

32. Van Voris, *Carrie Chapman Catt*, 133–134.

33. Ibid., 109.

34. Angela K. Smith, *Suffrage Discourse in Britain during the First World War* (Ashgate, 2005), 5, 23.

35. Jill Diane Zahniser and Amelia R. Fry, *Alice Paul: Claiming Power* (Oxford University Press, 2014), 95, 97.

36. Ibid., 87, 95.

37. Catt, *Woman Suffrage and Politics*, 242; Keyssar, *Right to Vote*, 167.

38. Paul quoted in Lynn Dumenil, *The Second Line of Defense: The Modern American Woman and World War I* (University of North Carolina Press, 2017), chap. 1.

39. Beth A. Behn, "Woodrow Wilson's Conversion Experience: The President and the Federal Woman Suffrage Amendment" (PhD diss., University of Amherst, 2012), 23.

40. Katherine Adams and Michael Keene, *Alice Paul and the American Suffrage Campaign* (University of Illinois, 2008), 30, 77, 201.

41. Wilson in Rodgers, *Atlantic Crossings*, 144; Wolloch, *Women and the American Experience*, 220.

CHAPTER TWO · NEUTRALITY DEFEATED, AND THE
TELEPHONE IN WAR AND PEACE

1. Larry Zuckerman, *The Rape of Belgium: The Untold Story of World War I* (New York University Press, 2004), 41.

2. Laurence Lafore, *The Long Fuse: An Interpretation of the Origins of World War I*, 2nd ed. (Waveland, 1997), 38.

3. Frederick A. Hale, "Fritz Fischer and the Historiography of World War One," *History Teacher* 9, no. 2 (February 1976): 258–279.

4. Isabel Hull, *A Scrap of Paper: Breaking and Making International Law during the Great War* (Cornell University Press, 2014), 42; Leslie Midkiff DeBauche, *Reel Patriotism: The Movies and World War I* (University of Wisconsin Press, 1997), 58.

5. Barbara Tuchman, *The Guns of August: The Outbreak of World War I* (Ballantine, 1994), 8; John N. Horne and Alan Kramer, *German Atrocities, 1914: A History of Denial* (Yale University Press, 2001), 40–42, 50–53; Niall Ferguson, *The Pity of War* (Penguin, 1998), 228.

6. Wilson in Sean Dennis Cashman, *America in the Age of the Titans: The Progressive Era and World War I* (New York University Press, 1988), 461.

7. Susan Grayzel, *At Home and Under Fire: Air Raids and Culture in Britain from the Great War to the Blitz* (Cambridge University Press, 2012), 21, 53.

8. Norman Stone, *World War One: A Short History* (Penguin, 2007), 99; John Terraine, *White Heat: The New Warfare, 1914–1918* (Sidgwick & Jackson, 1982), 35.

9. Mitchell Yockelson, *Forty-Seven Days: How Pershing's Warriors Came of Age to Defeat the German Army in World War I* (New American Library, 2016), 20.

10. Willa Cather, *One of Ours* (Alfred A. Knopf, 1922), 172–173.

11. DeBauche, *Reel Patriotism*, 16, 26–27, 58.

12. "A List of Neutral Ships Sunk by the Germans: From August 8th, 1914, to April 26th, 1917," National Archives of Australia, NAA: A5954, 192/34, http://recordsearch.naa.gov.au/SearchNRetrieve /Interface/ViewImage.aspx?B=645339&S=1.

13. George W. Baer, *One Hundred Years of Sea Power: The U.S. Navy, 1890–1990* (Stanford University Press, 1993), 67–68.

14. Ibid., 69.

15. Elizabeth Cobbs Hoffman, *American Umpire* (Harvard University Press, 2013), 189, 197. See also Wilson's War Message to Congress, 2 April 1917, http://www.presidency.ucsb.edu/ws/index .php?pid=65366.

16. Rodney P. Carlisle, *World War I* (Facts on File, 2007), 201–202.

17. *CR*, 65th Cong., 1st sess., 4 April 1917, 223.

18. Ibid., 187.

19. Ibid., 203, 222.

20. Ibid., 221; "Congress: Silver-Tongued Sunbeam," *Time*, 7 August 1939; James W. Johnson, *Arizona Politicians: The Noble and the Notorious* (University of Arizona Press, 2002), 114.

21. "List of Neutral Ships Sunk," 6; John Thompson, *A Sense of Power: The Roots of America's Global Role* (Cornell University Press, 2015), 86.

22. Cather, *One of Ours*, 173. See also Christopher Capozzola, *Uncle Sam Wants You: World War I and the Making of the Modern American Citizen* (Oxford University Press, 2008), 18.

23. *Report of the Chief Signal Officer to the Secretary of War, 1919* (GPO, 1919), 7 (hereafter cited as *CSO Annual Report*, 1919). Federal police protection began to emerge with the 1910 Mann Act, making the "white slave trade" illegal across state lines.

24. Alexis de Tocqueville, *Democracy in America*, trans. Henry Reeve (G. Adlard, 1839), 1:186; Gary Gerstle, *Liberty and Coercion: The Paradox of American Government from the Founding to the Present* (Princeton University Press, 2015), 6. See also Brian Balogh, *A Government Out of Sight: The Mystery of National Authority in Nineteenth Century America* (Cambridge University Press, 2009), 353, and *The Associational State: American Governance in the Twentieth Century* (University of Pennsylvania Press, 2015), 3; Capozzola, *Uncle Sam Wants You*, 7.

25. "Louvain to Honor America: Will Name New Streets for Washington, Wilson, and the Nation," *NYT*, 10 March 1915, 2; Bruno Cabanes, *The Great War and the Origins of Humanitarianism: 1918–1924* (Cambridge University Press, 2014), 207; Gerstle, *Liberty and Coercion*, 143.

26. "Bell Employees Asked to Aid in Food Conservation," *Bell Telephone News* 7, no. 2 (September 1917): 22.

27. Lettie Gavin, *American Women in World War I: They Also Served* (University Press of Colorado, 1997), 44; Julia Irwin, *Making the World Safe: The American Red Cross and a Nation's Humanitarian Awakening* (Oxford University Press, 2013), 12, 67.

28. Dumenil, *Second Line of Defense,* chap. 2; Kimberley Jensen, *Mobilizing Minerva: American Women in the First World War* (University of Illinois Press, 2008), 99.

29. *CR,* 65th Cong., 1st sess., 4 April 1917, 196.

30. Abraham L. Lavine, *Circuits of Victory* (Doubleday, 1921).

31. Sébastien Japrisot, *A Very Long Engagement,* trans. Linda Coverdale (Vintage, 2005), 26, 39.

32. *Radiotelegraphy (U.S. Signal Corps),* October 1916, 85–86, 93–127, http://earlyradiohistory.us/1916sc.htm.

33. Quoted in Betsy Rohaly Smoot, "Pioneers of U.S. Military Cryptology: Colonel Parker Hitt and His Wife Genevieve Parker Hitt," *Federal History* 4 (January 2012): 94.

34. Terraine, *White Heat,* 148.

35. M. D. Fagan, ed., *A History of Engineering and Science in the Bell System: The Early Years, 1875–1925* (Bell Telephone Laboratories, 1975), 471, 477–478.

36. Ibid., 56.

37. Anton Huurdeman, *The Worldwide History of Telecommunications* (Wiley & Sons, 2003), 237.

38. Ibid., 232.

39. Ibid., 316; Richard John, *Network Nation: Inventing American Telecommunications* (Harvard University Press, 2010), 234.

40. *Telephones and Telegraphs: 1902,* Bureau of the Census (GPO, 1906), 50–51.

41. Kenneth Lipartito, "When Women Were Switches: Technology, Work, and Gender in the Telephone Industry, 1890–1920," *American Historical Review* 99, no. 4 (October 1994): 1084–1085; John, *Network Nation,* 385.

42. Huurdeman, *Worldwide History of Telecommunications,* 194, 248, 484, 549–552, 608–609.

43. Lipartito, "When Women Were Switches," 1096; John, *Network Nation,* 231, 384; Huurdeman, *Worldwide History of Telecommunications,* 316.

44. Dorothy M. Johnson, "Confessions of a Telephone Girl," *Montana: The Magazine of Western History* 47, no. 4 (Winter 1997): 71, 74.

45. Huurdeman, *Worldwide History of Telecommunications*, 231; Lipartito, "When Women Were Switches," 1087.

46. Johnson, "Confessions of a Telephone Girl," 70. On factory wages, see Rede Davis, *California Women: A Guide to Their Politics, 1885–1911* (California Scene, c. 1968), 11–12.

47. Ellen Arguimbau, "From Party Lines and Barbed Wire: A History of Telephones in Montana," *Montana: The Magazine of Western History* 63, no. 3 (Autumn 2013): 41.

48. "A Portable Telephone," *Bell Telephone News* 7, no. 4 (November 1917): 7; Huurdeman, *Worldwide History of Telecommunications*, 231–233.

49. Huurdeman, *Worldwide History of Telecommunications*, 229–230.

50. Lavine, *Circuits of Victory*, 118; John J. Pershing, "General Pershing's Official Report of Battles Fought by American Forces in France from Their Organization to the Fall of Sedan," Report to the Secretary of War, 20 November 1918, Hoover Institution Library, Stanford University, 9.

51. John, *Network Nation*, 390–391.

52. Lavine, *Circuits of Victory*, 21.

53. Merle Egan Anderson to John de Butts, Chairman of the Board of AT&T, 26 November 1974, p. 1, NWWIM.

54. "The Army's Forgotten Women, from the Memoirs of Merle Egan Anderson," unpublished ms., Herron, 58 (hereafter cited as Egan Memoir); Thomas D. Snyder, ed., *Digest of Education Statistics, 1999*, U.S. Department of Education (GPO, 2000), 17.

55. Egan quoted in Michelle Christides, "Women and the Great War: A Story of the Generation Gap," unpublished ms., 1975, NWWIM, 32 (hereafter cited as Christides, "Women and the Great War").

56. "Miss Merle Egan, Operator," Fifth Unit, NARA, RG 111, Entry 45, Box 399.

CHAPTER THREE · LOOKING FOR SOLDIERS AND
FINDING WOMEN

1. Frederick Palmer, *John J. Pershing: General of the Armies* (Military Service Publishing, 1948), 124, 131; Nell Painter, *Standing*

at Armageddon: The United States, 1877–1919 (Norton, 1987), 331; *United States Army in the World War, 1917–1919*, vol. 1, *Organization of the American Expeditionary Forces* (Center of Military History, 1988), xv.

2. *Report of the Chief Signal Officer to the Secretary of War, 1918* (GPO, 1918), 14–15; Lavine, *Circuits of Victory*, 87–88, 123.

3. Lavine, *Circuits of Victory*, 56.

4. Ibid., 38–40; Palmer, *John J. Pershing*, 98.

5. Lavine, *Circuits of Victory*, 57, 68.

6. Jamie Stengle, "ATT's Historic Sculpture 'Golden Boy' Moves from NY to Dallas," *Lubbock Avalanche-Journal*, 28 July 2009, http://lubbockonline.com/stories/072809/sta_471635777.shtml#.Vhgc6Namdg0; Lavine, *Circuits of Victory*, 62.

7. Clarence Clendenen, *Blood on the Border: The United States Army and the Mexican Irregulars* (Macmillan, 1969), 315.

8. Squier quoted in Lavine, *Circuits of Victory*, 63; *CSO Annual Report*, 1919, 8.

9. John, *Network Nation*, 393, 400.

10. Lavine, *Circuits of Victory*, 70.

11. Ibid., 74–75.

12. AT&T employee A. Abraham Lavine served as captain in the Army Air Service during the war, and later wrote the official history of ATT's involvement. Lavine, *Circuits of Victory*, 98, 124; "Extra Pay in Signal Corps," *Bell Telephone News* 7, no. 4 (November 1917): 16.

13. French staff officer quoted in Dan Carlin, *Blueprint for Armageddon IV*, podcast audio, Hardcore History, episode 53, 54:48, accessed 13 October 2015, http://www.dancarlin.com/hardcore-history-53-blueprint-for-armageddon-iv/.

14. Lavine, *Circuits of Victory*, 115–120.

15. Ibid., 125.

16. Egan Memoir, 27.

17. Gavin, *American Women in World War I*, 2.

18. Lee A. Craig, *Josephus Daniels: His Life and Times* (University of North Carolina Press, 2013), 12–14, 86, 190, 365–366, 380–381;

Donald Mathews and Jane Sherron De Hart, *Sex, Gender, and the Politics of ERA: A State and the Nation* (Oxford University Press, 1990), 233.

19. Gavin, *American Women in World War I*, 2–3, 5.

20. Ibid., 3.

21. Pearl James, ed., *Picture This: World War I Posters and Visual Culture* (University of Nebraska, 2010), 290; Hoover Institution Archives Poster Collection, accessed 31 October 2015, http://www .politicalposters.hoover.org/find?keys=Gee!!+I+wish+I+were+a+ma n&country=All.

22. Gavin, *American Women in World War I*, 6, 9.

23. Ibid., 5.

24. "Yeoman Sonnenberg, U.S.N.," *Bell Telephone News* 7, no. 6 (January 1918): 23.

25. Memorandum for the Assistant Chief of Staff, "Utilization of Women in the Military Services," Anita Phipps, 6 August 1926, 5; NARA, RG 165, G-1, Box 9, Numerical File 1921–1942, 6853 to 7000; Linda Hewitt, *Women Marines in World War I* (GPO, 1974); Hoover Institution Archives Poster Collection, accessed 1 October 2015, http://www.politicalposters.hoover.org/find? keys=if +you+want+to+fight,+join+the+marines&country=All.

26. Lucy Noakes, *Women in the British Army: War and the Gentle Sex, 1907–1948* (Routledge, 2006), 62. See also Joanna Bourke, "Women on the Home Front in World War One," 3 March 2011, http://www.bbc.co.uk/history/british/britain_wwone/women _employment_01.shtml.

27. Christabel Pankhurst quoted by Linda Ford, "Alice Paul and the Politics of Nonviolent Protest," in Baker, *Votes for Women*, 180; Noakes, *Women in the British Army*, 47; Patricia Greenwood Harrison, *Connecting Links: The British and American Woman Suffrage Movements, 1900–1914* (Greenwood, 2000), 208.

28. Baker, *Votes for Women*, 148.

29. Noakes, *Women in the British Army*, 48.

30. Ibid., 59; Grayzel, *At Home and Under Fire*, 5.

31. Noakes, *Women in the British Army*, 66, 69, 72.

32. Ibid., 81; British Imperial War Museum, Catalog no. Art.IWM PST 13167, accessed 2 April 2017, http://www.iwm.org.uk/collec tions/item/object/31371.

33. "Military Sphere of Women Widens," *Macon (GA) Telegraph*, 3 September 1917, 2.

34. Ute Daniel, *The War from Within: German Working-Class Women in the First World War* (Oxford, 1997).

35. Beatty quoted in Melissa Stockdale, "'My Death for the Motherland Is Happiness': Women, Patriotism, and Soldiering in Russia's Great War, 1914–1917," *American Historical Review* 109, no. 1 (February 2004): 96 (also in Jensen, *Mobilizing Minerva*, 60–61).

36. "The Battalion of Death," *Bell Telephone News* 7, no. 2 (September 1917): 23.

37. I. W. Littell, Colonel, Quartermaster Corps, to Officer in Charge of Cantonment Construction, 17 September 1917. NARA, RG 111, Entry 45, Box 396, File: Telephone Operators, 1917–1919.

38. "Memorandum for the Chief of Staff, Regarding Female Employees, Telephone and Laundry," 12 November 1917, 1, NARA, RG 165, G-1, Box 9, Numerical File 1921–1942, 6853 to 7000.

39. M. G. Spinks, Hq. Coast Artillery Training Center to Commanding General, 3rd Corps Area, Baltimore, 14 October 1920, NARA, RG 111, Entry 45, Box 397.

40. Major D. G. Beatty, Signal Corps, to the Quartermaster General, 22 October 1917, Subject: Furnishings for Use of Civilian Telephone Operators. Brigadier General I. W. Littell, Officer in Charge of Cantonment Construction to Chief Signal Officer, War Department, 2 November 1917, NARA, RG 111, Entry 45, Box 396.

41. "Memorandum from the Adjutant General to the Commanding Generals of All National Guard and National Army Cantonments, Subject: Female Employees," 11 December 1917, 1. See also Mattie Treadwell, *The Women's Army Corps,* United States Army in World War II (GPO, 1953), 7.

42. Nancy Bristow, *Making Men Moral: Social Engineering during the Great War* (New York University, 1996), 2.

43. Ibid., 135.

44. Memorandum, G. R. Johnston to Captain Louis M. Evans, 23 February 1918, 1–2, NARA College Park, RG 111, Entry 45, Box 396.

45. Ibid.

46. Palmer, *John J. Pershing*, 130, 155.

47. R. B. Owens to Colonel C. McK. Saltzman, Office of the Chief Signal Office, 20 February 1920, NARA, RG 111, Entry 45, Box 398; *CSO Annual Report*, 1919, 16.

48. Quoted in Ruth Quinn, "An Army Wife 'Doing Her Bit' in World War I: The Story of Genevieve Young Hitt," 31 March 2014, http://www.army.mil/article/122935/An_Army_Wife__Doing_Her_Bit__in_World_War_I__The_Story_of_Genevieve_Young_Hitt/.

49. Parker Hitt to "George" [presumably Squier], 28 February 1931, NARA, RG 111, Entry 45, Box 398.

50. *CSO Annual Report*, 1919, 5, 7, 9.

51. "Women in the Armed Forces," *Armed Forces Talk*, 7 November 1952, attached to the "Statement of Mark Hough in Support of Enactment of S. 1414, Before the Committee on Veterans' Affairs, 25 May 1977," MEA.

52. David Kennedy, *Over Here: The First World War and American Society* (Oxford University Press, 1980), 171.

53. Frank Vandiver, *Black Jack: The Life and Times of John J. Pershing* (Texas A&M University Press, 1977), 1:14, 171.

54. Susan Zeiger, *In Uncle Sam's Service: Women Workers with the American Expeditionary Force, 1917–1919* (University of Pennsylvania Press, 2004), 17.

55. General John Pershing, Headquarters, AEF, to Washington, 8 November 1917, NARA, RG 111, Entry 45, Box 396, File: Telephone Operators, 1917–1919; Norwood, *Labor's Flaming Youth*, 32.

56. General John Pershing, Headquarters, AEF, to Washington, 8 November 1917, NARA, RG 111, Entry 45, Box 396, File: Telephone Operators, 1917–1919; Palmer, *John J. Pershing*, 131.

57. John J. Pershing, *General Pershing's Story of the American Army in France* (Herzig & McLean, 1919), 35.

CHAPTER FOUR · WE'RE GOING OVER

1. Ernest J. Wessen, Midland Book Company, to Mrs. H. R. Anderson, 14 August 1950, 1, NWWIM.

2. Ibid.

3. E. J. Wessen to the Office of the CSO, 14 January 1918. Nellie F. Snow, DOB 2-12-1881, NPRC.

4. The National Archives contain many of the letters Wessen sent between mid-November and early December 1917. See, for example, E. J. Wessen, Office of Chief Signal Officer, to Editor, *New Orleans Times-Picayune*, 10 November 1917, NARA, RG 111, Entry 45, Box 396.

5. Ernest J. Wessen, Midland Book Company, to Mrs. H. R. Anderson, 14, 22, and 30 August 1950, 1, NWWIM.

6. R. F. Estabrook, "Report of Interview, Applicants for Position with Signal Corps Telephone Operator's Unit, General Information on Which Applicants Should Be Informed," undated, c. 1918, ATT-SA, 176-03-01-06.

7. P. G. Harris, Adjutant General to Chief Signal Officer, 28 November 1917; George O. Squier to the Adjutant General, 7 March 1919, reproduced in "A Study of the Service of Woman Telephone Switchboard Operators of the A.E.F.," 27 April 1926, 13, NARA, RG 111, Entry 45, Boxes 398 and 400.

8. J.A.G.O. to the Adjutant General, 7 March 1918, Regarding Telephone Operators, 231.37, sent to the Chief Signal Officer on 9 March 1918, NARA, RG 111, Entry 45, Box 400.

9. Stuart G. Shepard, Treasury Department Bureau of War Risk Insurance to the Chief Signal Officer, 20 March 1918, reproduced in "A Study of the Service of Woman Telephone Switchboard Operators of the A.E.F.," 27 April 1926, 10, NARA, RG 111, Entry 45, Box 400.

10. See certificate of war risk insurance issued to Olive Shaw, 23 March 1918, Affidavit of Olive Shaw, 1 May 1977, attached to the "Statement of Mark Hough in Support of Enactment of S. 1414, MEA.

11. "Synopsis of Action Taken on Suggested Legislation for the Enlistment of Women in the Military Service in the World War," 2–3, NARA, RG 165, G-1, Box 9, Numerical File 1921–1942, 6853 to 7000.

12. Ibid. For a fuller account of the doctors' fight, see Jensen, *Mobilizing Minerva*, 77–97.

13. Baker quoted in Palmer, *John J. Pershing*, 83, 203.

14. Wessen quoted in "American Telephone Girls at the Front," *Southern Telephone News* 6, no. 7 (July 1918): 19, ATT-SA; "U.S. Hello Girls Success in War," *Oregonian*, 17 May 1918, 4, and *University of Washington Daily*, 14 May 1918, cited in Jill Frahm, "The Hello Girls: Women Telephone Operators with the American Expeditionary Forces during World War I," *Journal of the Gilded Age and Progressive Era* 3, no. 3 (July 2004): 274.

15. See Wessen to Mrs. H. R. [Merle] Anderson, 14 August 1950, 2, NWWIM.

16. "Suffrage Scores with Club Women, Solicitor Baker and Other Says Equality Is Coming," *Cleveland Plain Dealer*, 19 January 1911, 3.

17. Frederick Palmer, *Newton D. Baker* (Dodd, Mead, 1931), 2:34–35.

18. Newton Baker to Chairman, House Committee on Military Affairs, 26 December 1917, cited in Treadwell, *Women's Army Corps*, 8.

19. Palmer, *Newton D. Baker*, 2:61, 67; Robert Ferrell, *America's Deadliest Battle: Meuse-Argonne, 1918* (University of Kansas Press, 2007), 12.

20. *Times-Picayune*, 17 November 1917, 9; *Kansas City Star*, 5 December 1917, 1; *Philadelphia Inquirer*, 30 December 1917, 8; *Duluth News-Tribune*, 6 December 1917, 2.

21. "Young Women of America, Attention!" *Bell Telephone News* 7, no. 7 (February 1918): 23.

22. "Bell Operator Enlists in Army," *Southwestern Telephone News*, March 1918, 68, ATT-SA.

23. Lavine, *Circuits of Victory*, 272, 275; "American Telephone Girls at the Front," 18.

24. E. J. Wessen, "Information Relative to Telephone Operators for Duty in France," 11 December 1917, Affidavit of Gertrude Hoppock, "Statement of Mark Hough in Support of Enactment of S. 1414," MEA.

25. R. F. Estabrook to Captain W. S. Vivian, 11 June 1918, reproduced in "A Study of the Service of Woman Telephone Switchboard

Operators of the A.E.F.," 27 April 1926, 7, NARA, RG 111, Entry 45, Box 400; "Our Telephone Girls in France," 16 June 1918, John J. Carty Papers, Box 82-02-02, ATT-NJ.

26. "Request for 100 French-Speaking Operators for Service Abroad," undated, c. December 1917, Carty Papers, ATT-NJ.

27. Frahm, "The Hello Girls," 276.

28. Memoir of Beatrice Pauline Bourneuf-Savard, 1, Herron.

29. "What the War Means," *Bell Telephone News* 7, no. 7 (February 1918): 8; Painter, *Standing at Armageddon*, 330.

30. The National Archives no longer possesses the records of every member of the women's Signal Corps. Of the records the author was able to obtain, only Grace Banker requested this consideration. Grace Banker, NPRC.

31. "The Church Club," *Barnard Bulletin*, 10 January 1912, 1; "Church Club Meeting," *Barnard Bulletin*, 8 May 1912, 2, Online Archives, Barnard Library.

32. "Sophomore Show," *Barnard Bulletin*, 20 November 1912, 1–2, online archives, Barnard Library.

33. Grace Banker to Chief Signal Officer (hereafter cited as CSO), 9 December 1917; Banker to CSO, 7 January 1917, Grace Banker, NPRC. On Banker's position at AT&T before the war, see "Introduction, Part I, Louise Barbour, Unit Five," 1, Louise Barbour Papers, Schlesinger Library, Harvard University, MC 219, Signal Corps Questionnaire and Barbour Article.

34. Louise LeBreton, interview by Matt Herron, c. 1980, Herron.

35. Louise and Raymonde LeBreton, NPRC; affidavit of Louise LeBreton, 1 April 1977, 2, attached to "Statement of Mark Hough in Support of Enactment of S. 1414," MEA.

36. Louise and Raymonde LeBreton, NPRC; Louise LeBreton to Room 806, 6 December 1917, 1; Quotes from Official Telegrams to R. F. Estabrook, 12 January 1918 and 14 January 1918. See also Matt Herron, "Hello Central! Give Me No Man's Land," unpublished ms., 22 August 1988, 5, Herron.

37. Louise LeBreton Maxwell, "My Experiences as a Telephone Operator in France, 1918–1919," 1, Herron.

38. LeBreton, interview by Herron.

39. See four handwritten letters from Hope Kervin to CSO, the War Department, 7–23 December 1917; telegram of 5 January 1918; and her "Application for Certificate in Lieu of Lost or Destroyed Discharge Certificate," Margaret H. [Hope] Kervin, NPRC.

40. Captain William S. Vivian to Lieutenant E. J. Wessen, 28 December 1917, NARA, RG 111, Entry 45, Box 398.

41. See, for example, Martha Patterson, *Beyond the Gibson Girl: Reimagining the American New Woman, 1895–1915* (University of Illinois Press, 2005).

42. "Telephone Operators Join Our Army," *Telephone Review,* March 1918, 72, ATT-SA.

43. Personnel record of Madeleine Batta, 19 December 1893, NPRC; LeBreton quoted in Christides, "Women and the Great War," 32.

44. Isabelle Villiers to E. J. Wessen, 18 December 1917; R. F. Estabrook to Ernest J. Wessen, 4 January 1918; Isabelle Villiers (25 August 1894), NPRC.

45. Frahm, "The Hello Girls," 278, 292; "Confidential Report of Interviewer," 25 January 1918, Marie A. Gagnon (17 February 1894), NPRC.

46. "To Whom It May Concern," Frank Miracle, National War Savings Committee, 11 July 1918; Merle Egan (26 July 1988), NPRC; "The County's Roll of Honor" and "Helena War Savings Societies Buy a Total of $7,497 Thrifties in June," *Helena Daily Independent,* 4 July 1918, 6, and 23 July 1918, 2.

47. Grace Banker, NPRC.

48. "Signal Corps Days in the A.E.F., 1918–1919." Typewritten Transcription of the Diary of Grace Banker Paddock, 1, in the papers of her son-in-law Robert Timbie, Exeter, New Hampshire (hereafter cited as Banker Diary).

49. President, California Packing Corporation to Chief Signal Officer, 19 December 1917; Inez Crittenden, NPRC.

50. Inez Crittenden, NPRC. See War Risk Insurance, Melina J. Adam, NPRC.

51. See, for example, Evelyn LaRiviere, Anita Brown, Beryl Broderick, Cordelia Dupuis, and Nellie Snow, NPRC.

52. R. F. Estabrook, "Report of Interview, Applicants for Position with Signal Corps Telephone Operator's Unit, General Information on Which Applicants Should Be Informed," undated, c. 1918, 6, ATT-SA, 176-03-01-06; "U.S. Hello Girls Success in War," *Portland Oregonian*, 17 May 1918, 4; "Battalions of American War Girls Operating Telephone Exchanges behind Front Lines in France," *Greensboro Daily Record*, 17 May 1918, 7.

53. "Miss Blanche Louise Barbour, Operator," NARA, RG 111, Entry 45, Box 399, Fifth Unit; Wilburt Ranney to Mr. Darling, ATT, 19 October 1918, Alice Langelier, NPRC.

54. Christides, "Women and the Great War," 46; Merle E. Anderson to R. E. Lynch, 2 August 1977, NWWIM.

55. Helen Ruth Orb (b. November 29, 1894), NPRC.

56. Quoted in Frahm, "The Hello Girls," 288; Minnie Goldman, NPRC.

57. "Miss Minnie Hermine Goldman, Operator," 2, 5, NARA, RG 111, Entry 45, Box 399, Sixth Unit.

58. Constance Ball to D. D. Field, New York Telephone Company, 16 March 1918, Lucienne Bigou (9 September 1889), NPRC.

59. "Confidential Report of Interviewer," 1 January 1918, Cordelia Dupuis, NPRC; "To Whom It May Concern," Frank Miracle, National War Savings Committee, 11 July 1918, and "Confidential Report of Interviewer," 11 July 1918, Merle Egan, NPRC. See also report on Merle Egan in NARA, RG 111, Entry 45, Box 399, Fifth Unit; Gavin, *American Women in World War I*, 80.

CHAPTER FIVE · PACK YOUR KIT

1. Merle Egan, NPRC.

2. Lavine, *Circuits of Victory*, 274.

3. Ibid., 275.

4. Janet R. Jones, NPRC.

5. Lavine, *Circuits of Victory*, 277.

6. Anita Brown, 9 September 1889, NPRC.

7. "Miss Marie Josephine Hess, Operator," NARA, RG 111, Entry 45, Box 399, Seventh Unit; Marie Josephine Hess, NPRC.

8. R. F. Estabrook, "Report of Interview, Applicants for Position with Signal Corps Telephone Operator's Unit, General Information on Which Applicants Should Be Informed," undated, c. 1918, 3, ATT-SA, 176-03-01-06.

9. "Eleanor Hoppock Follows Sister," *University of Washington Daily*, 22 March 1918, NWWIM; Frahm, "The Hello Girls," 285, 291.

10. "Co-Ed Will Serve U.S. as Hello Girl," *University of Washington Daily*, 15 February 1918, NWWIM.

11. Adele Hoppock to "Dearest Muz," 25 February–3 March 1918; "Two Sons and a Daughter of A. A. McKillop Wearing Uniform," *Seattle Post-Intelligencer*, undated, c. September 1918, NWWIM.

12. Adele Hoppock to "Dearest Puz" and "Dearest Eleanor," 5–8 March 1918, NWWIM.

13. "Diary of Berthe M. Hunt, a Woman Who Served in the U.S. Army in France," 1918, typed transcription by Matt Herron (hereafter cited as Hunt Diary), 2–3, NWWIM; Lucille Crane, interview by Matt Herron, transcript, 5, 11, 16, Herron. On married women, see Frahm, "The Hello Girls," 283.

14. "War Department, Office of the Chief Signal Officer," 2 October 1918, "Information Regarding Signal Corps Telephone Operator's Unit," ATT-SA, 176-03-01-06.

15. Hunt Diary, 2.

16. Frahm, "The Hello Girls," 292.

17. "Operators for France," *Pacific Telephone Magazine*, April 1918, 8, ATT-SA.

18. Hunt Diary, 3–4.

19. Eleanor Hoppock to "Dearest Family," 30 March 1918, and to "Dearest Mother," 1 April 1918, NWWIM.

20. Eleanor Hoppock to "Dear Mother," 17 May 1918, and to "Dearest Family," 5 June 1918, NWWIM.

21. "Information Regarding Signal Corps Telephone Operator's Unit," War Department, Office of the Chief Signal Officer, 2 October 1918, 1, ATT-SA, 176-03-01-06; Hunt Diary, 5.

22. "Telephone Operators Join Our Army," *Telephone Review*, March 1918, 71, 73.

23. Hunt Diary, 5; Adele Hoppock to Eleanor, 20 April 1918, 1, NWWIM.

24. Adele Hoppock to Eleanor Hoppock, 20 April 1918, 3; Louise LeBreton Maxwell, "My Experiences," 2, NWWIM.

25. Christides, "Women and the Great War," 13.

26. R. F. Estabrook, "Report of Interview, Applicants for Position with Signal Corps Telephone Operator's Unit, General Information on Which Applicants Should Be Informed," undated, c. 1918, ATT-SA, 176-03-01-06.

27. Grace Banker, "My Experience as an Operator in France," 1, AHC, File: Grace Banker (Paddock).

28. Adele Hoppock to "Dear Mother," 16 April 1918, 1–2, NWWIM.

29. Ibid.

30. Inez Haynes Irwin, *The Californiacs* (A. M. Robertson, 1916), 2; Hunt Diary, 6.

31. Hunt Diary, 4–5.

32. Ibid., 4; 1st Lieut. Eugene D. Hill to CSO, 27 September 1918, NARA, RG 111, Entry 45, Box 399, Fifth Unit.

33. Blanche-Yvonne Kelly to "Gentlemen, ATT," 22 August 1936, 1, NARA, RG 111, Entry 45, Box 398.

34. Paula S. Fass, *The Damned and the Beautiful: American Youth in the 1920s* (Oxford University Press, 1970), 260–268; R. L. Estabrook to E. J. Wessen, 15 June 1918, 1, NARA, RG 111, Entry 45, Box 400.

35. R. F. Estabrook to Ernest J. Wessen, 26 March 1918, 1–3, NARA, RG 111, Entry 45, Box 400.

36. Ibid.

37. Ibid.

38. E. J. Wessen to R. F. Estabrook, 30 March 1918, 1–2, NARA, RG 111, Entry 45, Box 400; Wessen to Adjutant General of the Army, 8 April 1918, Subject "Telephone Unit," NARA, RG 111, Entry 45, Box 398.

39. E. J. Wessen to R. F. Estabrook, 30 March 1918, 1, NARA, RG 111, Entry 45, Box 400.

40. Hunt Diary, 5. Wilson quoted in "Roll of Honor, Employees of the Receivers, Central Union Telephone Company," *Bell Telephone News*, February 1918.

41. Adele Hoppock to Eleanor, 20 April 1918, NWWIM; Egan Memoir, 39.

42. Hunt Diary, 6.

43. Adele Hoppock to "Dearest Muz and Puz," 16 April 1918, NWWIM.

44. Zeiger, *In Uncle Sam's Service*, 11–12.

45. Banker Diary, 1.

46. Ibid.

CHAPTER SIX · WILSON ADOPTS SUFFRAGE, AND THE SIGNAL CORPS EMBARKS

1. "President Puts Suffrage to Fore," *NYT*, 26 October 1917, 1, 24.

2. Ibid.

3. Beth A. Behn, "Woodrow Wilson's Conversion Experience: The President and the Federal Woman Suffrage Amendment," diss. 2012, University of Amherst, 181–182.

4. For a historiographical overview, see Behn, "Woodrow Wilson's Conversion Experience," 1–13; William O'Neill, *Everyone Was Brave: The Rise and Fall of Feminism in America* (Quadrangle, 1969); Eleanor Flexner, *Century of Struggle: The Woman's Rights Movement in the United States* (Harvard University Press, 1959); Christine Lunardini, *From Equal Suffrage to Equal Rights: Alice Paul and the National Women's Party, 1910–1928* (New York University Press, 1986).

5. Behn, "Woodrow Wilson's Conversion Experience," 220.

6. Victoria Bissell Brown, "Did Woodrow Wilson's Gender Politics Matter?" in *Reconsidering Woodrow Wilson*, ed. John Milton Cooper Jr. (Woodrow Wilson Center, 2008), 129, 134.

7. Rosalyn Terborg-Penn, *African-American Women in the Struggle for the Vote, 1850–1920* (Indiana University Press, 1998), 122–123.

8. Michael Perman, *Pursuit of Unity: A Political History of the American South* (University of North Carolina Press, 2009), 215.

9. "Wilson Backs Amendment for Woman Suffrage," *NYT,* 10 January 1918, 1, 3.

10. Weyl quoted in Daniel Rodgers, *Atlantic Crossings: Social Politics in a Progressive Age* (Harvard University Press, 1998), 41–42, 75.

11. Cobbs Hoffman, *American Umpire,* 206.

12. "Wilson's Fourteen Points," accessed 31 October 2015, https://wwi.lib.byu.edu/index.php/President_Wilson%27s _Fourteen_Points.

13. Behn, "Woodrow Wilson's Conversion Experience," 219.

14. *CR,* 65th Cong., 2nd sess., 10 January 1918, 771, 773.

15. Ibid., 764–765.

16. Ibid., 767.

17. "House for Suffrage, 274 to 136, Exact Two-Thirds Vote Required," *NYT,* 11 January 1918, 1.

18. Catt, *Woman Suffrage and Politics,* 322.

19. Brown, "Did Woodrow Wilson's Gender Politics Matter?" 129.

20. *NYT,* 10 January 1918, 3.

21. R. H. Gibson and Maurice Prendergast, *The German Submarine War: 1914–1918* (1931; Periscope, 2002), 146, 152.

22. Banker Diary, 1.

23. Baer, *One Hundred Years of Sea Power,* 73.

24. Hunt Diary, 8–9.

25. Egan Memoir, 10–11.

26. Benedict Crowell, *How America Went to War: The Road to France,* 1 (Yale University Press, 1921), 104; Palmer, *Newton D. Baker* (Dodd, Mead, 1931), 2:364.

27. Christides, "Women and the Great War," 47.

28. Alison Rowe and Cheryl Regehr, "Whatever Gets You through Today: An Examination of Cynical Humor among Emergency

Service Professionals," *Journal of Loss and Trauma* 15, no. 5 (September / October 2010): 449–450.

29. Agnes T. Houley, NPRC; Christides, "Women and the Great War," 50.

30. Banker Diary, 4.

31. Robert M. Grant, *U-Boats Destroyed: The Effect of Anti-Submarine Warfare, 1914–1918* (Putnam, 1964), 91.

32. Hunt Diary, 9; Adele Hoppock Mills, "Signal Corps Reminiscences," 2, Herron; Grant, *U-Boats Destroyed*, 84, 89–90, 118.

33. Baer, *One Hundred Years of Sea Power*, 76.

34. Banker Diary, 3; *Historical Statistics*, 2:1140; Affidavit of Louise LeBreton Maxwell, 3, attached to "Statement of Mark Hough in Support of Enactment of S. 1414," MEA.

35. Alfred W. Crosby, *America's Forgotten Pandemic: The Influenza of 1918* (Cambridge University Press, 2003), 5–6.

36. Ibid., 123.

37. *History of the USS Leviathan: Cruiser and Transport Services, United States Atlantic Fleet* (Brooklyn Eagle Press, 1919), 92, https://archive.org/stream/historyofusslevi00broo/history ofusslevi00broo_djvu.txt.

38. Jane Conroy, NPRC.

39. Christides, "Women and the Great War," 48.

40. Jane Conroy, NPRC.

41. Commanding Officer, U.S. Troops, On Board SS *Carmania* to The Commanding General, 77th Division, 11 April 1918, 2, NWWIM.

42. "Report of Disinterment, Preparation, Shipment and Reburial of Body," 25 August 1921, 1–2; Veteran's Burial Record, Inez Crittenden, NPRC. See also Irene Givenwilson to George Squier, 20 November 1922, NARA, RG 111, Entry 45, Box 398.

43. H. C. Houlthan, Compensation and Insurance Claims Section, to Adjutant General, 5 March 1919, 1; Inez Crittenden, Civilian Record, NPRC.

44. Hunt Diary, 7.

45. Banker Diary, 2–3.

46. Palmer, *John J. Pershing*, 149–150.

47. Palmer, *Newton D. Baker*, 2:4.

48. Banker Diary, 3.

49. Ibid., 2, 3.

50. Egan in Christides, "Women and the Great War," 13; Adele to Eleanor Hoppock, 20 April 1918, 3, NWWIM.

51. Egan Memoir, 5–6.

52. Ibid., 15–16.

53. Banker Diary, 5, 7; Hunt Diary, 10.

54. Banker Diary, 10.

55. Kennedy, *Over Here*, 173.

56. Terraine, *White Heat*, 278.

57. Stone, *World War One*, 139; Michael Howard, *The First World War* (Oxford University Press, 2002), 116.

58. David Trask, *The AEF and Coalition Warmaking: 1917–1918* (University of Kansas Press, 1993), 43.

59. Michael Neiberg, *Fighting the Great War: A Global History* (Harvard University Press, 2005), 314, 319.

60. Terraine, *White Heat*, 291; William Shirer, *Collapse of the Third Republic: An Inquiry into the Fall of France in 1940* (Simon and Schuster, 1969), 132.

61. Hunt Diary, 10.

62. Ellen Turner to "Dear Family," 23 August 1918, 2, NWWIM.

63. Banker Diary, 12.

64. Louise LeBreton Maxwell, "My Experiences as a Telephone Operator," 4, Herron.

65. Dupuis quoted in Gavin, *American Women in World War I*, 81.

66. Banker Diary, 13.

67. "Young Women Christian Association of the War Work Program of the Industrial Committee," Chicago, 10 September 1918, 17, YWCA, Box 702, Folder 4; Captain A. L. Hart, Paris Signal Officer,

in "A Study of the Service of Woman Telephone Operators, of the A.E.F.," 21, NARA, RG 111, Entry 45, Box 400.

68. Terraine, *White Heat*, 148–149.

CHAPTER SEVEN · AMERICANS FIND THEIR WAY, OVER THERE

1. Philippe Bernard and Henri Dubief, *The Decline of the Third Republic, 1914–1929* (Cambridge University Press, 1985), 79–80.

2. Martin van Creveld, *Technology and War* (Free Press, 1991), 175.

3. Graves quoted in Crosby, *America's Forgotten Pandemic*, 17.

4. Richard W. Stewart, ed., "The U.S. Army in World War I, 1917–1918," chap. 1 in *Army Military History*, vol. 2, *The United States Army in a Global Era, 1917–2003* (U.S. Army Center of Military History, 2005), 42, http://www.history.army.mil/books /amh-v2/PDF/Chapter01.pdf.

5. Creveld, *Technology and War*, 189; John J. Pershing, *My Experiences in the World War*, (Frederick Stokes, 1931), 1:160.

6. Quoted in Kennedy, *Over Here*, 192. On the critique of Wilson, see Robert Ferrell, *America's Deadliest Battle: Meuse-Argonne, 1918* (University Press of Kansas, 2007), 1–17.

7. Stewart, "U.S. Army in World War I."

8. Trask, *AEF and Coalition Warmaking*, 53.

9. Cooper quoted in Frank Friedel, *Over There: The Story of America's First Great Overseas Crusade* (McGraw-Hill, 1990), 102.

10. *CSO Annual Report*, 1919, 163, 714.

11. "To the Sixth Telephone Operating Unit," p. 1, Signal Corps AEF, Headquarters SOS, Hough Papers.

12. Yockelson, *Forty-Seven Days*, 12.

13. Hunt Diary, 16; Banker Diary, 28.

14. Hindy Lauer Schachter, *Frederick Taylor and the Public Administration Community: A Reevaluation* (State University of New York Press, 1989), 19–20; Robert Kanigel, *The One Best Way: Frederick Winslow Taylor and the Enigma of Efficiency* (MIT Press, 1997), 41.

15. Norwood, *Labor's Flaming Youth*, 33–40.

16. "Telephone and Telegraph Division, Bulletins n. 1 and 2, Military Telephone Regulations and Telephone Department Instructions," Chief Signal Office, AEF, 1 September 1918 (reissued), Section 2, 1–2, Herron (hereafter cited as *Telephone Regulations*). See also *CSO Report, 1919,* 172.

17. Hunt Diary, 16, 20–21.

18. Harry Williams, "'Hello' in Paris," *Los Angeles Times,* 14 May 1918, 114.

19. Hunt Diary, 12, 16, 22, 38.

20. Adele Hoppock Mills, "Signal Corps Reminiscences," 3, Herron.

21. Grace Banker, "My Experiences as a Signal Corp Operator in France," 2–3, AHC; Hunt Diary, 28.

22. Grace Banker, "My Experiences as a Signal Corp Operator in France," 3; *Annual Report, Signal Corps Work in France, 1918,* YWCA, Box 7, Folder 16, 13.

23. *Telephone Regulations,* sec. 2, pp. 1–4.

24. Ibid., sec. 2, p. 4, sec. 4, pp. 1–2, 5; Banker Diary, 30.

25. Hunt Diary, 19, 22, 37.

26. Grace Banker, "I Was a Hello Girl," *Yankee Magazine,* AHC, File: Grace Banker Paddock, 71.

27. *CSO Report, 1919,* 162, 173–174, 179.

28. Ibid., 1919, 179.

29. Ibid., 1919, 448.

30. Banker Diary, 19.

31. Egan Memoir, 38.

32. Banker Diary, 29.

33. Louise LeBreton Maxwell, "My Experiences as a Telephone Operator in France, 1918–1919," 8–9, Herron.

34. Banker Diary, 18, 23, 28.

35. Ibid., 39.

36. Yockelson, *Forty-Seven Days,* 12.

37. Mrs. John Wehrely [Honey Fay], "The Hello Girls: Pluggers of World War I," *Detroit Sunday Magazine,* 7 November 1976.

38. *Historical Statistics,* 2:1141. In 1916, the U.S. Army numbered 108,339 men; in 1918, it numbered 2,395,742.

39. Gavin, *American Women in World War I,* 184, 187, 214–215.

40. "Supplementary Report of Mary George White for March 1918," 1, YWCA, Box 707, Folder 13.

41. "The Arrival of the Telephone Girls," *YWCA Overseas,* 21, YWCA, Box 707, Folder 6; "Signal Corps Telephone Operators' Billets," Headquarters, S.O.S., 1 November 1918, 3, YWCA, Box 7, Folder 16.

42. *Annual Report, Signal Corps Work in France, 1918,* YWCA, Box 7, Folder 16, 21.

43. Ibid., 13, 18.

44. Ibid., 8–9.

45. Ibid., 15.

46. Ibid., 5.

47. Egan Memoir, 30; Hunt Diary, 10.

48. Banker Diary, 27, 36.

49. "Young Women's Christian Association War Work Program of the Industrial Committee: Portions of the Address by Miss Henrietta Roelofs," Chicago, 10 September 1918, 16–18, YWCA, Box 702, Folder 4.

50. *Annual Report, Signal Corps Work in France, 1918,* 5, 8, 14, YWCA, Box 7, Folder 16.

51. Ibid., 8, 14, 21.

52. "US 'Hello Girls' in War Work Make Big Hit in Paris," *Philadelphia Inquirer,* 29 September 1918, 1.

53. Banker Diary, 43.

54. *Annual Report, Signal Corps Work in France, 1918,* 13, YWCA, Box 7, Folder 16.

55. Parker Hitt to Genevieve Hitt, 21 April 1918, Parker Hitt (Private Papers), courtesy of David and Evie Moreman and Kevin and Jennifer Mustain.

56. Banker Diary, 23.

57. Ibid., 38.

58. Ellen Turner to "Dear Family," 23 August 1918, 1, NWWIM; Banker Diary, 44, 46, 59.

59. Egan Memoir, 8, 74, 78, 121.

60. Harbord to Chief Signal Officer, 5 January 1919, SOS Cablegram, Lucienne Bigou, NPRC.

61. *Annual Report, Signal Corps Work in France, 1918*, 16, YWCA, Box 7, Folder 16.

62. Hunt Diary 22, 33, 34.

63. Ibid., 29.

64. Ibid., 32, 38.

65. Ibid., 21–22, 28.

66. Ibid., 24, 26.

67. Egan Memoir, 16, 19.

68. Hunt Diary, 42.

69. Banker Diary, 18, 35, 46.

70. Gavin, *American Women in World War I*, 44, 46, 51; *Annual Report, Signal Corps Work in France, 1918*, 15, YWCA, Box 7, Folder 16.

71. Commanding Officer, U.S. Troops, On Board SS *Carmania* to the Commanding General, 77th Division, 11 April 1918, 1, NWWIM.

72. Hunt Diary, 19.

73. Banker Diary, 22, 24, 27, 32.

74. Ibid., 24.

75. Post Signal Officer J. W. Riser to Miss Grace Banker, Chief Operator, Headquarters, American Expeditionary Forces, Subject: Violation of Censorship Rules, 21 May 1918, 1, Herron; Banker Diary, 45.

76. "Signal Corps Reminiscences," Adele Hoppock Mills, 5, Herron.

77. Louise LeBreton Maxwell, "My Experiences as a Telephone Operator in France," 10–11.

78. Ibid., 3; Banker Diary, 14.

79. Yockelson, *Forty-Seven Days*, 143, 284.

CHAPTER EIGHT · BETTER LATE THAN NEVER
ON THE MARNE

1. *United States Army in the World War, 1917–1919*, vol. 1, *Organization of the American Expeditionary Forces* (GPO: 1988), xxv.

2. Pershing, *My Experiences*, 2:194.

3. Egan Memoir, 28–29.

4. Ibid., 29.

5. Ellen Turner to "Dear Family," 23 August 1918, 4–5; "War Letters," Adele Hoppock, 16 May 1918, *Pacific Telephone Magazine*, clipping, NWWIM.

6. Yockelson, *Forty-Seven Days*, 220–228.

7. "US Hello Girls in War Work Make Big Hit in Paris," *Philadelphia Inquirer*, 29 September 1918, 1.

8. Hunt Diary, 14.

9. Pershing, *My Experiences*, 2:54.

10. Lavine, *Circuits of Victory*, 359; "Hello Girls Under Fire," *Duluth News-Tribune*, 15 December 1918, 14, 116; Isaac Marcosson, *S.O.S.: America's Miracle in France* (John Lane, 1919), 116.

11. Banker Diary, 35.

12. Friedel, *Over There*, 108; Martin Marix Evans, ed., *American Voices of World War I: Primary Source Documents, 1917–1920* (Fitzroy Dearborn, 2001), 74.

13. Hunt Diary, 14.

14. Evans, *American Voices of World War I*, 76–77.

15. Pershing, *My Experiences*, 2:60, 285; Yockelson, *Forty-Seven Days*, 48.

16. Neiberg, *Fighting the Great War*, 329.

17. Hunt Diary, 15.

18. Ibid., 16.

19. Harbord in Friedel, *Over There,* 112.

20. Evans, *American Voices of World War I,* 83.

21. Ibid., 85.

22. *CR,* 65th Cong., 3rd sess. (GPO, 1919), 2701; Neiberg, *Fighting the Great War,* 328.

23. Hunt Diary, 26, 27.

24. Banker Diary, 48.

25. Quoted in Byron Farwell, *Over There: The United States in the Great War, 1917–1918* (Norton, 1999), 183.

26. Banker Diary, 50.

27. Yockelson, *Forty-Seven Days,* 45.

28. Parker Hitt, 13 November 1918, in "A Study of the Service of Woman Telephone Operators of the A.E.F.," 27 April 1926, 23, NARA, RG 111, Entry 45, Box 400.

29. Banker Diary, 52.

30. Ibid., 51, 66, 70; Hunt Diary, 18; Ralph D. Paine, "Our Navy and the French Coast," *Saturday Evening Post,* 20 April 1918, 14.

31. Hunt Diary, 35; Banker Diary, 52.

32. Ibid.; Ibid.

33. Neiberg, *Fighting the Great War,* 333. See attachments to Anita Phipps, "Memorandum for the Assistant Chief of Staff, Utilization of Women in the Military Service," 7 January 1927 (hereafter cited as Phipps Memorandum): General C. C. Williams to the Adjutant General of the Army, 23 August 1918, and Williams to War Plans Division, 28 September 1918, NARA, RG 165, G-1, Box 9, Numerical File 1921–1942, 6853 to 7000. See General Harbord's Cable, and Colonel Ira Reeves to Assistant Secretary of War, 24 August 1918, 5, in "Synopsis of Action Taken on Utilization of Women," 3, appendix to Phipps Memorandum.

34. "Synopsis of Action," 3–4, appendix to Phipps Memorandum.

35. Ibid., 5–6, appendix to Phipps Memorandum.

36. Ibid., 6, appendix to Phipps Memorandum.

37. Louise Barbour to "Darling Mother," 24 November 1918, Louise Barbour Papers, Schlesinger Library, Harvard University, MC 219, Folder 1.

38. Hunt Diary, 36, 37.

39. Banker Diary, 55; Hunt Diary, 36.

40. Hunt Diary, 36. Parker Hitt's papers contain letters from Suzanne Prevot after the war, indicating their camaraderie. Prevot to Hitt, 9 October 1920, Parker Hitt (Private Papers), courtesy of David and Evie Moreman and Kevin and Jennifer Mustain.

41. Hunt Diary, 37.

42. Pershing, *My Experiences*, 2:246.

43. Ibid.

44. Ibid., 2:254–255.

45. Hunt Diary, 39.

46. Ibid., 37–39.

47. Banker Diary, 56–57.

48. Report of Colonel Parker Hitt, 1, NARA-CP, RG 111, Entry 45, Box 400.

49. Banker Diary, 58.

50. Alan Palmer, *Victory* (Grove, 1998), 209; Yockelson, *Forty-Seven Days*, 77.

51. Hunt Diary, 40; "Real Thrills in This Operator's War," *Telephone Bulletin*, New England Telephone Company, March 1919, 1, ATT-SA.

52. Banker Diary, 58.

53. Ibid.; Hunt Diary, 40.

54. Hunt Diary, 40–41.

55. Ibid., 41; Banker Diary, 59.

56. Hunt Diary, 41.

57. Pershing, *My Experiences*, 1:175.

58. Hunt Diary, 41; "Real Thrills in This Operator's War," 2, ATT-SA.

59. Cablegram No. 1707, General Pershing, 20 September 1918, in Memorandum from Office of the Chief Signal Officer to Adjutant General of the Army, 24 September 1920, NARA, RG 111, Entry 45, Box 398.

CHAPTER NINE · WILSON FIGHTS FOR DEMOCRACY AT HOME

1. Robert Dallek, "Woodrow Wilson, Politician," *Wilson Quarterly* 15, no. 4 (Autumn 1991): 106–114.

2. Roosevelt in John Milton Cooper, *Breaking the Heart of the World: Woodrow Wilson and the Fight for the League of Nations* (Cambridge University Press, 2001), 30.

3. Cobbs Hoffman, *American Umpire*, 199–200.

4. Roosevelt in Kennedy, *Over Here*, 235.

5. Wilson to Wolcott, 9 May 1918, in Behn, "Woodrow Wilson's Conversion Experience," 228.

6. Behn, "Woodrow Wilson's Conversion Experience," 229–230.

7. "Wilson Spurs Fight for Women's Vote," *NYT,* 14 June 1918, 10.

8. Behn, "Woodrow Wilson's Conversion Experience," 232.

9. Wilson to Shields, 20 June 1918, in Behn, "Woodrow Wilson's Conversion Experience," 232.

10. Behn, "Woodrow Wilson's Conversion Experience," 238.

11. "Suffrage Summary: Mrs. Jessie Wilson Sayre Asked Soldiers to Fight for Suffrage," *Woman Citizen* 1, no. 25 (November 1917): 477.

12. William McAdoo, *Crowded Years* (Houghton Mifflin, 1931), 497.

13. *CR,* 65th Cong., 2nd sess. (GPO, 1918), 10844.

14. Ibid., 10843, 10854.

15. Ibid., 10778–10779.

16. Ibid., 10779.

17. Ibid., 10779, 10783.

18. Ibid., 10775.

19. Ibid., 10771, 10772, 10894.

20. Ira Katznelson, *Fear Itself: The New Deal and the Origins of Our Time* (Norton, 2003), 8.

21. *CR*, 65th Cong., 2nd sess. (GPO, 1918), 10931.

22. Ibid., 10773.

23. Ibid.

24. Dumenil, *Second Line of Defense*, introduction.

25. *CR*, 65th Cong., 2nd sess. (GPO, 1918), 10773–10774; Anne Wiltsher, *Most Dangerous Women: Feminist Peace Campaigners of the Great War* (Pandora, 1985), 83; C. Roland Marchand, *The American Peace Movement and Social Reform, 1889–1918* (Princeton University Press, 1972), 187.

26. George Mosse, *The Image of Man: The Creation of Modern Masculinity* (Oxford University Press, 1996), 53; Stockdale, "'My Death for the Motherland Is Happiness,'" 82.

27. *CR*, 65th Cong., 2nd sess. (GPO, 1918), 10786.

28. Ibid., 10898–10899.

29. Ibid., 10781.

30. Ibid., 10775.

31. Mary Dudziak, *Cold War Civil Rights: Race and the Image of American Democracy* (Princeton University Press, 2000).

32. *CR*, 65th Cong., 2nd sess. (GPO, 1918), 10927.

33. Ibid., 10892.

34. "Gives Joy to Suffragists," *NYT*, 1 October 1918, 13.

35. McAdoo, *Crowded Years*, 498.

36. *CR*, 65th Cong., 2nd sess. (GPO, 1918), 10928–10929.

37. "Wilson Makes Suffrage Appeal, but Senate Waits," *NYT*, 1 October 1918, 1, 13.

38. *CR*, 65th Cong., 2nd sess. (GPO, 1918), 10982.

39. Behn, "Woodrow Wilson's Conversion Experience," 241; *CR*, 65th Cong., 2nd sess. (GPO, 1918), 10929, 10985.

40. Wilson quoted in Behn, "Woodrow Wilson's Conversion Experience," 242.

41. "Woman's Party in Fight: Works for Defeat of Democratic Candidates for Senate," *NYT,* 4 November 1918, 10; Kennedy, *Over Here,* 241–243.

42. "Women Cast Votes as Readily as Men," *NYT,* 6 November 1918, 4.

43. Kennedy, *Over Here,* 237.

CHAPTER TEN · TOGETHER IN THE CRISIS OF MEUSE-ARGONNE

1. Hunt Diary, 41–42.

2. Ibid., 42–43.

3. Banker Diary, 62. Hunt recorded the incident as well, noting that shrapnel had fallen four feet away. Hunt Diary, 42–43.

4. Robert Ferrell, *America's Deadliest Battle: Meuse-Argonne, 1918* (University Press of Kansas, 2007), 37.

5. Colonel George Crile quoted in Farrell, *America's Deadliest Battle,* 39.

6. "A Study of the Service of Woman Telephone Operators of the A.E.F.," 27 April 1926, 25, NARA, RG 111, Entry 45, Box 400.

7. Hunt Diary, 43; Farrell, *America's Deadliest Battle,* 41; Adele Hoppock Mills, "Signal Corps Reminiscences," 5, Herron.

8. Eddie Rickenbacker, *Fighting the Flying Circus* (Frederick Stokes, 1919), 269; Yockelson, *Forty-Seven Days,* 102.

9. Banker Diary, 63.

10. Ibid.

11. Ibid., 76–77.

12. Adele Hoppock to Mother, 5 October 1918, 10, NWWIM.

13. Ibid., 8, NWWIM.

14. Ibid.; Chad Williams, *Torchbearers of Democracy: African American Soldiers in the World War I Era* (University of North Carolina Press, 2010), 119–120, 136–139; Pershing, *My Experiences,* 2:45–46; Gavin, *American Women in World War I,* 60.

15. Hunt quoted in Lavine, *Circuits of Victory,* 565; Hunt Diary, 46, NWWIM.

16. Adele Hoppock to Mother, 5 October 1918, 9, NWWIM.

17. Banker Diary, 65.

18. Ibid. Farrell, *America's Deadliest Battle*, 54–55.

19. Hunt Diary, 44, 46; Banker Diary, 72; Adele Hoppock to Mother, 5 October 1918, 1, NWWIM; Banker Diary, 72.

20. Banker Diary, 70–71.

21. Ibid., 69–71, 75, 79, 81–82.

22. Ibid., 72, 76; Pershing, *My Experiences*, 2:327.

23. Banker Diary, 62, 72–73; Farrell, *America's Deadliest Battle*, 122.

24. Farrell, *America's Deadliest Battle*, 76, 84; Liggett in Yockelson, *Forty-Seven Days*, 80; York in Michael Neiberg, *Second Battle of the Marne* (Indiana University Press, 2008), 43.

25. George C. Marshall, *Memoirs of My Service in the World War, 1919–1918* (Houghton Mifflin, 1976), 208; Yockelson, *Forty-Seven Days*, 99.

26. Farrell, *America's Deadliest Battle*, 105, 155.

27. Richard Slotkin, *Lost Battalions: The Great War and the Crisis of American Nationality* (Holt, 2005), 263; Farrell, *America's Deadliest Battle*, 77.

28. Slotkin, *Lost Battalions*, 333; Lavine, *Circuits of Victory*, 569; Hunt Diary, 46.

29. Adele Hoppock to Mother, 5 October 1918, 2, NWWIM.

30. "Six Hello Girls Help First Army," *Lexington Leader*, 25 October 1918, 3.

31. Banker Diary, 64, 75; Hunt Diary, 48.

32. Adele Hoppock to "Dearest Mother and Father," 1 November 1918, 6, NWWIM.

33. Hunt Diary, 48.

34. Ibid., 38, 44.

35. Ibid., 49.

36. Ibid., 49–50; Banker Diary, 80–81.

37. Banker Diary, 86.

38. Ibid., 89.

39. Hunt Diary, 48.

40. Yockelson, *Forty-Seven Days*, 218, 242.

41. Ibid., 242; Hunt Diary, 48.

42. Hunt Diary, 51; Pershing, *My Experiences*, 2:348–349; Rickenbacker, *Fighting the Flying Circus*, 160.

43. Banker Diary, 72, 87.

44. Adele Hoppock to Mother, 5 October 1918, 4, NWWIM.

45. Banker Diary, 87.

46. Ibid., 87.

47. Ibid., 89.

48. Ibid., 88.

49. Parker Hitt, 13 November 1918, in "A Study of the Service of Woman Telephone Operators of the A.E.F.," 27 April 1926, 23, NARA, RG 111, Entry 45, Box 400; Hunt Diary, 52.

50. Hunt Diary, 52.

51. Ibid., 52.

52. Banker Diary, 89, 92.

53. Marshall, 194.

54. Ibid.

55. Trask, *AEF and Coalition Warmaking*, 177.

56. Hunt Diary, 54.

57. Evans, *American Voices of World War I*, 160; Banker Diary, 91; Nurse Laura Frost in Gavin, *American Women in World War I*, 61.

58. Banker Diary, 90.

59. Thompson, *Sense of Power*, 105.

60. Marjorie Kinnan, "The Blue Triangle Follows the Switchboard," *Pacific Telephone Magazine*, March 1919, 16.

CHAPTER ELEVEN · PEACE WITHOUT THEIR VICTORY MEDALS

1. Egan Memoir, 53; Hunt Diary, 55.

2. Christides, "Women and the Great War," 34.

3. Behn, "Woodrow Wilson's Conversion Experience," 246.

4. "Address Fails to Stir: Democrats Almost Alone in Applauding Wilson's Trip," *NYT*, 3 December 1918, 1; Cooper, *Breaking the Heart*, 35.

5. Woodrow Wilson, "Sixth Annual Message," December 2, 1918, American Presidency Project, http://www.presidency.ucsb.edu/ws/index.php?pid=29559.

6. Ibid.

7. Suffrage Success Held to Be Assured," *NYT*, 9 December 1918, 22.

8. Julia Mickenberg, "Suffragettes and Soviets: American Feminists and the Specter of Revolutionary Russia," *Journal of American History* (March 2012): 1033.

9. "14 Women Seek Seats in Commons," *NYT*, 9 December 1918, 3; Catt, *Woman Suffrage and Politics*, 337.

10. Wilson quoted in Behn, "Woodrow Wilson's Conversion Experience," 247; "Suffragists Burn Wilson in Effigy; Many Locked Up," *NYT*, 10 February 1919, 1.

11. *CR*, 65th Cong., 3rd sess. (GPO, 1919), 3054, 3056.

12. "Militants Demand a Special Session," *NYT*, 11 March 1919, 10.

13. "Six Suffragettes Put Under Arrest" and "Senators Bitter after Lodge Move," *NYT*, 5 March 1919, 3, 4; Cooper, *Breaking the Heart*, 68.

14. Joseph Tumulty, *Woodrow Wilson as I know Him* (Doubleday, 1921), 423–425.

15. Lodge quoted in William Widenor, *Henry Cabot Lodge and the Search for an American Foreign Policy* (University of California Press, 1980), 9.

16. Lodge quoted in Cooper, *Breaking the Heart*, 51.

17. Cooper, *Breaking the Heart*, 92.

18. Ibid., 55, 93.

19. *CR*, 66th Cong., 1st sess. (GPO, 1919), 42.

20. Ibid., 85, 89, 570; Paul Ortiz, *Emancipation Betrayed: The Hidden History of Black Organizing and White Violence in Florida from Reconstruction to the Bloody Election of 1920* (University of California Press, 2005), 161–162.

21. *CR*, 66th Cong., 1st sess., 83.

22. Ibid., 80.

23. Ibid., 82, 88; Winthrop, "A Model of Christian Charity," *Winthrop Papers*, vol. 2, *1623–1630* (Massachusetts Historical Society, 1931), 295.

24. *CR*, 66th Cong., 1st sess., 82–84.

25. "Suffrage Wins Easily in House," *NYT*, 22 May 1919, 1.

26. *CR*, 66th Cong., 1st sess., 564.

27. Ibid., 617–619.

28. Ibid., 625, 627.

29. Ibid., 620.

30. "Wilson Suffrage Plea," *NYT*, 3 September 1919, 10; "Urges North Carolina to Ratify Suffrage," *NYT*, 26 June 1920, 6.

31. Catt, *Woman Suffrage and Politics*, 489–490.

32. Philips quoted in Catt, *Woman Suffrage and Politics*, 62.

33. Catt, *Woman Suffrage and Politics*, 107–108.

34. Anna Harvey, *Votes without Leverage: Women in American Electoral Politics, 1920–1970* (Cambridge University Press, 1998), 6, 167.

35. Harvey, *Votes without Leverage*, 1.

36. Terborg-Penn, *African-American Women in the Struggle for the Vote*.

37. Cooper, *Breaking the Heart*, 47, 65, 119.

38. "E. Russel to the Members of the Telephone Operating Unit, Signal Corps, AEF," 12 November 1918, in *Memento of the Telephone Operating Units*, NWWIM, copy of Anna Campbell.

39. Ibid.

40. Ibid.

41. Parker Hitt, "A Study of the Service of Woman Telephone Operators of the A.E.F.," 27 April 1926, 26, NARA, RG 111, Entry 45, Box 400.

42. LeBreton Personnel File, NPRC; Affidavit of Alma Hawkins, 13 May 1977, 4, attached to the "Statement of Mark Hough in Support of Enactment of S. 1414," MEA; "A Study of the Service of Woman Telephone Operators of the A.E.F.," 27 April 1926, 28, NARA, RG 111, Entry 45, Box 400.

43. Grace Banker, NPRC; "Hello Girl, Who Was Given D.S.M., Says All Girls Were Heroes," *Canton (OH) Repository,* 19 September 1919, 4.

44. "First Woman to Enlist in U.S. Army Gets D.S.M. for Heroism during War," *Denver Post,* 6 June 1919, 27; *Augusta Chronicle,* 26 September 1919, 10. See similar articles in *Philadelphia Inquirer,* 26 May 1919; *Kansas City Star,* 26 May 1919, 9; *Kansas City Star,* 2 June 1919, 13; *Fort Wayne New-Sentinel,* 13 June 1919; *Evansville Courier,* 14 June 1919; *Duluth News-Tribune,* 21 September 1919; and *Pueblo Chieftain,* 21 September 1919, 8. Grace Banker, NPRC.

45. "Hello Girl, Who Was Given D.S.M., Says All Girls Were Heroes," 4.

46. *CSO Annual Report,* 1919, 525.

47. Bullard, Telegram received at GHQ Toul, 14 February 1919, US Signal Corps, NWWIM.

48. Egan Memoir, 128.

49. Hunt Diary, 56–58.

50. Ibid., 59–61.

51. "Annual Report, Signal Corps Work in France, 1918 (dated 26 February 1919)," 23, YWCA, Box 707, Folder 16.

52. Hunt Diary, 57, 60, 62, 64–66.

53. "My Experiences as a Chief Operator in France," Grace Banker Paddock, November 1937, 7–8, MEA.

54. Egan Memoir, 102–103, 129; Office of the Chief Signal Officer, War Department, to Zone Finance Office, 27 December 1919, 1, NARA-CP, RG 111, Entry 45, Box 396.

55. "Grave Location Blank," Burial Record of Corah Bartlett, NPRC; "Michigan Heroine Is Dead in France," c. 1919, AHC, File: Cora

Bartlett; "Bartlett Funeral Impressive Affair," *Jackson City Patriot*, 1 February 1922, 3. (Name inconsistent in records.)

56. Burial Record of Corah Bartlett, NPRC; Photos of "Military Funeral of Cora Bartlett at Tours, France," *Bell Telephone News* 9, no. 5 (December 1919), Milwaukee, Wisconsin, ATT-SA.

57. Special Orders No. 224, Headquarters Base Section, No. 5, 12 August 1919, 1; Affidavit of Gertrude Hoppock, 13 May 1977, 4, attached to "Statement of Mark Hough in Support of Enactment of S. 1414," MEA.

CHAPTER TWELVE · SOLDIERING FORWARD IN THE TWENTIETH CENTURY

1. Bickford to Squier, 12 November 1920, Christine Bickford, NPRC. Kervin to Adjutant General, Washington, D.C., Stamp-dated 22 June 1928, Margaret H. Kervin, NPRC. Bonus Application, 1936, Louise Chaix, NPRC. C.O., Fitzsimmons General [Army] Hospital, Colorado, 6 October 1921, Irene Gifford, NPRC.

2. "Seattle Troops Arrive in New York From France," *Seattle Daily Times*, 3 June 1919, 13.

3. Egan Memoir, 151–152.

4. "Annual Report, Signal Corps Work in France, 1918 (26 February 1919)," 24. YWCA, Box 707, Folder 16.

5. Egan Memoir, 50, 64, 102, 167. "Miss Merle Egan Will have Charge of Central at Versailles Palace," *Anaconda Standard*, 4 January 1919, 12. "Montana Girl Picked for Important Post at Paris," *Dixon Herald*, 13 January 1919.

6. Egan Memoir, 165, 168.

7. Egan Memoir, 179.

8. Edgar Russel (CSO, AEF) to George Squier (CSO), 23 January 1919, 1. R. F. Estabrook to Lt. Col. John Moore, 13 December 1918, 2, NARA, RG 111, Entry 45, Box 398.

9. Squier to Adjutant General, 12 February 1919, in "A Study of the Service of Woman Telephone Operators, of the AEF," 12, RG 111, Entry 45, Box 400.

10. LiPartito, 1087.

11. Jos. Trany to CSO Squier, 25 February 1919, in "A Study of the Service of Woman Telephone Operators, of the AEF," 13, RG 111, Entry 45, Box 400.

12. Squier to Trany, 7 March 1919, and Trany to Squier, 21 March 1919, in "A Study of the Service of Woman Telephone Operators, of the A.E.F.," 13–14, 18, RG 111, Entry 45, Box 400.

13. "Synopsis of Action Taken on Utilization of Women with the Army in the World War," Point 19, NARA, RG 165, G-1, Box 9, Numerical File 1921–1942, 6853 to 7000.

14. L. W. Layton, Assistant Traffic Supervisor, to Signal Officer, War Department, 14 June 1919, 2, NARA, RG 111, Entry 45, Box 398.

15. "To the Secretary of War of the United States," attached to correspondence from Jos. Haas to Senator Medill McCormick, 13 May 1919, NARA, RG 111, Entry 45, Box 398.

16. Saltzman to McCormick, 1 July 1919, 2, NARA, RG 111, Entry 45, Box 398.

17. Martin Morrison to the Secretary of War, 25 November 1919, 1, RG 111, Entry 45, Box 398.

18. G. I. Jones for the Surgeon General, 13 November 1919, 1, Lt. Col. John Moore, Memorandum for Civilian Personnel Branch, 1 December 1919, 2, NARA, RG 111, Entry 45, Box 398.

19. R. B. Owens to Col. C. Mck. Saltzman, 20 February 1920, NARA, RG 111, Entry 45, Box 398.

20. Testimony of Mrs. Ray W. Flanery, 3, in "Extract of Hearings Before Committee on War Veterans Legislation," H.R. 69th Congress, 1928, NARA, RG 165, G-1, Box 9, Numerical File 1921–1942, 6853 to 7000.

21. George Squier to Secretary of War, 25 May 1922, NARA, RG 111, Entry 45, Box 400; F. R. Curtis, Memorandum for the Adjutant General of the Army, 30 January 1920, 2, NARA, RG 111, Entry 45, Box 400. On the service of Quartermaster and Ordnance women, see "Statement of Mary Norman," 6, in "Extract of Hearings Before Committee on War Veterans Legislation," H.R. 69th Congress, 1928, NARA, RG 165, G-1, Box 9, Numerical File 1921–1942, 6853 to 7000.

22. Phipps Memorandum, 2.

23. Roy H. Coles to State Bonus Boards, Office of the Chief Signal Officer, Headquarters 2D Corp Area, 17 May 1921, Herron. See Coles to Anna Campbell, 17 June 1921, NWWIM.

24. Wessen to Merle Egan Anderson, 14 August 1950, 2, NWWIM.

25. Stephen Walmsley to L. R. Gignilliat, 24 February 24 1921; Irving J. Carr to Edward Moran, Jr., 3 January 1934, NARA, RG 111, Entry 45, Box 398.

26. E. T. Conley to Adele Louise [Hoppock] Mills, 14 May 1935, 1; "Arguments by Olive Shaw Who Worked for Edith Nourse Rogers to be Presented to Armed Services Comm., 1950," 1, NWWIM.

27. Conley to Anna Maude McMullen, 22 June 1936, 1. A. Maude McMullen, NPRC.

28. Anderson to Cyrus Vance, 26 June 1953; "Women's Lib—1918," attached to letter from Merle Anderson to Esther Fitts, 3 January 1975, NWWIM.

29. "Overseas Club of Women Asks Congressional Charter," *Denver Times*, 12 March 1922, WOSL Scrapbooks, UTSA, Special Collections, Box 222, MS-22; *A History of the Women's Overseas Service League*, Helene Sillia, July 1978, 5, 73, UTSA, Box 5, File 3.

30. Anderson to "Dear Friend and Buddy," 8 December 1930, NWWIM.

31. Mabel Hooper to Eleanor [Hoppock], 17 February 1942, NWWIM.

32. Grace Banker to "Miss Smith," 16 November 1937, 1, letter in personal possession of Carolyn Timbie, Andover, NH.

33. Leisa D. Meyer, *Creating G.I. Jane: Sexuality and Power in the Women's Army Corps during World War II* (Columbia University Press, 1996), 16, 78; Margaret Bakker to Unit Presidents, 4 February 1943, WOSL, UTSA, Box 2, File 6.

34. Wessen to Anderson, 14 August 1950, 2, and 2 August 1950, 1, NWWIM

35. "To Provide Military Status for Women Who Served Overseas with the Army of the U.S. during World War I (H.R. 471). Mr. Elston," 5 March 1947, Committee on Armed Services, Serial

66, H.R. 471, 80th Congress, 1st Session, 6 January 1947, 2; Patterson to Walter Andrews, 5 March 1947 (H.R. 471), 929–930, Sudoc: Y4.Ar5 / 2a:947–48 / 66.

36. *Congressional Record—Appendix*, 22 June 1950 (GPO: 1950), A4877.

37. "Arguments by Olive Shaw Who Worked for Edith Nourse Rogers to Be Presented to Armed Services Committee, 1950," NWWIM.

38. Olin Teague to Mrs. H. R. Anderson, 11 May 1960, NWWIM.

39. Edward Kennedy to Newbert, 28 March 1963; Humphrey to Newbert, 25 March 1963; Mansfield to Newbert, 25 March 1963; Dirksen to Newbert, 25 March 1963, Hough; Peter Garland to President Kennedy, 10 April 1961; Newbert to Merle Anderson, 29 April 1961, 2, NWWIM.

40. "Merle" to Harold Say, 9 March 1972, 1, NWWIM; Mrs. H. R. Anderson to Signal Corps "Buddies," 20 September 1967, Herron; Merle Anderson to Colonel John LaRe, 16 April 1974, Herron.

41. Lucille Crane, interview by Matt Herron, 28, Herron; Louise to Merle, 21 May 1975, Herron.

42. WOSL Roster, 19 October 1969, 1–2, UTSA, Box 16, File 5; *A History of the Women's Overseas Service League*, Helene Sillia, July 1978, 172, UTSA, Box 5, File 3.

43. Anderson to A. O. Soderholm, 6 August 1971; J. B. Koch, National Commander, to Anderson, 28 October 1971, Hough; Handwritten note to Louise LeBreton on letter to Edmund Norris from Mrs. H. R. Anderson, 18 September 1973, Herron.

44. Merle to Mrs. William Abbott (Hilda Van Brunt), 2 December 1976, NWWIM.

45. Christides, "Women and the Great War," 33.

46. "Merle" to "Louise," 25 March 1973, and Edmund Norris to Mrs. H. R. Anderson, 18 September 1973, Herron; "I Was a Hello Girl," Grace Banker Paddock, *Yankee*, March 1974.

47. Oliver Meadows to Merle Anderson, 21 August 1972, NWWIM.

48. Merle E. Anderson to Oliver Meadows," 26 August 1972, NWWIM.

49. G. Gavin Mackenzie and Robert Weisbrot, *The Liberal Hour: Washington and the Politics of Change in the 1960s* (Penguin, 2008), 162; Ellen Carol DuBois and Lynn Dumenil, *Through Women's Eyes: An American History* (Bedford / St. Martins, 2016), 608.

50. Nancy McLean, *Freedom Is Not Enough: The Opening of the American Workplace* (Harvard University Press, 2006), 125, 129.

51. Merle Anderson to General George Brown, 4 June 1974; Merle to Harold Say, 10 January 1972, NWWIM.

52. Charlotte Gyss Terry to "My Dear Mrs. Anderson," 9 November 1972 [*sic*, 1971], NWWIM; "Siberian Corps Ruled Army Vets," *Dallas Morning News*, 26 March 1971, D-3.

53. "It's Official," *Palm Beach Post*, 28 March 1971, A-5.

54. Merle A. to "Charlotte," 11 November 1971, 3, NWWIM.

55. Harold Say to Mrs. H. R. Anderson, 18 January 1972, NWWIM.

56. Anderson to Green, 23 January 1972, NWWIM.

57. "Harold" to "Merle," 21 April 1974, NWWIM; Edith Green to Merle Anderson, 17 May 1974, Herron.

58. "Merle" to "Louise," undated, c. June 1974 and 11 June 1974, Herron.

59. Anderson to Holt and Boggs, 10 October 1974, NWWIM.

60. Pat Leeper to Mrs. H. R. Anderson, 2 March 1975 and 15 May 1975, NWWIM.

61. Leeper to Anderson, 23 May 1975, NWWIM.

62. Anderson to Pat Leeper, 18 May 1975. NWWIM.

63. Leeper to Anderson, three undated letters c. May 1975 and September 1975, and 22 June 1975, NWWIM.

64. Pat McGivern Leeper, "Why I Am a Feminist," *Washington Times Magazine*, 2 July 1975, 25; Merle Anderson to John Greenwald, 7 July 1975, NWWIM.

65. *Seattle Times*, 20 July 1975; Mark Hough, interview by author, Bellevue, Washington, 1 September 2015.

66. Merle to Louise, 12 October 1976, 1–2, Herron.

67. Quote from "The Long Battle: A Curious Chapter in Armed Services History," *Columbus Dispatch*, 21 October 1973; Goldwater to Hough, 12 October 1976; Hough to Selma Samols, 18 January 1977; Hough to Harold Say, 10 February 1977, NWWIM.

68. Testimony of Barry Goldwater, *Recognition for Purposes of VA Benefits: Hearing Before the Committee on Veterans' Affairs*, 95th Cong., 1st sess. (GPO, 1977), 34, 37–38.

69. Merle to Mark, 14 November (year undated); Merle to Mark, 28 April 1977, NWWIM.

70. *Recognition for Purposes of VA Benefits: Hearing Before the Committee on Veterans' Affairs*, 95th Cong., 1st sess. (GPO, 1977), 139.

71. "Statement of Mark Hough," *Recognition for Purposes of VA Benefits: Hearing Before the Committee on Veterans' Affairs*, 95th Congr., 1st sess. (GPO, 1977), 307.

72. Elizabeth Bohan to Jimmy Carter, 15 June 1977; Bohan to "All Units," 25 February 1977, 1, NWWIM.

73. Mrs. Selby E. Pels to Merle, 15 March 1977, 2, NWWIM.

74. See comptroller's comments attached to letter from Senator Henry Jackson to Mark Hough, 27 April 1977, 4, NWWIM.

75. R. E. Lynch to Mrs. H. R. Anderson, 10 August 1977, NWWIM.

76. Hough to "Ladies of the Signal Corps and Their Friends," 10 November 1977; "Signal Corps Women Who Have Received Army Discharges," NWWIM.

77. "Army Says Goodbye to Hello Girls," *Seattle Post-Intelligencer*, 29 August 1979, A3.

78. Mary E. Birse to Mark Hough, 1 June 1979, 1–2, NWWIM.

EPILOGUE

1. Michelle Christides, "The History of a Hello Girl," accessed 18 April 2016, http://www.worldwar1.com/dbc/hello.htm.

2. "Female Pilot Unit Gains Support in Congress for Right to Arlington Burials," *NYT*, 27 February 2016, A12.

3. Cokie Roberts and Steven V. Roberts, "Draft Rosie!," 23 March 2016, http://www.uexpress.com/cokie-and-steven-roberts /2016/3/23/draft-rosie.

4. http://www.foxnews.com/us/2016/09/07/new-law-allows-female -wwii-pilot-to-be-inurned-at-arlington.html. Accessed 19 September 2016.

5. "Transcript: Donald Trump's Taped Comments About Women," *NYT*, 8 October 2016, http://www.nytimes.com/2016/10/08/us /donald-trump-tape-transcript.html; "Transcript: Michelle Obama's Speech on Donald Trump's Alleged Treatment of Women," *NPR*, 13 October 2016, http://www.npr.org/2016/10/13/497846667/transcript -michelle-obamas-speech-on-donald-trumps-alleged-treatment-of -women.

ACKNOWLEDGMENTS

I AM INDEBTED to Mark Hough, Matthew Herron, and the Timbie Family (Robert, Grace, and Carolyn) for sharing the unpublished private papers of Merle Egan Anderson, Berthe Hunt, Louise LeBreton Maxwell, Grace Banker Paddock, and other Signal Corps operators. I am grateful as well to Terry Fife and Elizabeth Trantowski of History Works in Chicago for locating Grace Timbie, Grace Banker's only daughter. They made this book possible.

Others smoothed my path. Frank Alva, Myra Burton, William Caughlin, Catherine Clinton, Robert Cobbs, Lynn Dumenil, Jill Frahm, Rita Gibson, Maida Goodwin, Sheldon Hochheiser, Holly Reed, Bruce Schulman, Betsy Rohaly Smoot, Lori Tagg, and Mitchell Yockelson contributed generously. Their knowledge guided me. Their kindness encouraged me.

The staff of the National Personnel Record Center in St. Louis, especially Cara Moore, provided invaluable assistance. I received support from Texas A&M University, San Diego State University, and Hoover Institution at Stanford. Chris Dauer of Hoover and David Vaught of Texas A&M were unfailingly helpful. Joyce Seltzer of Harvard University Press is all one could wish for in an editor. Kathleen Drummy and Bridget Martin of Harvard sped the project, assisted by Michele Mattingly, Mary Ribesky, and Jamie Nan Thaman. Publicist Gretchen Crary of February Media made me laugh and work harder.

Family and friends listened patiently when I told (and re-told) the story of the Hello Girls. They sweetened my adventure. The former telephone operators in my family deserve honorable mention: my aunt Diane Shelby Ewing and my sister-in-law Barbara Anger. Special thanks go to my children, Victoria and Gregory Shelby, and my husband, Jim Shelley.

Lastly, I honor the memory of Joyce Nower, professor and founder of America's first Women's Studies Program, who asked me at fifteen what I thought about feminism. Like the Hello Girls, she inspires me to this day.

INDEX